Transportation Statistics

Edited by Brian W. Sloboda

ISBN-13: 978-1-60427-008-2

Printed and bound in the U.S.A. Printed on acid-free paper

10 9 8 7 6 5 4 3 2 1

Library of Congress Cataloging-in-Publication Data
Transportation statistics / edited by Brian W. Sloboda.
p. cm.
Includes index.
ISBN 978-1-60427-008-2 (hbk. : alk. paper)
1. Transportation—Statistics. 2. Transportation—Statistical methods.
I. Sloboda, Brian W., 1968–
HE191.5.T73 2009
388.072′7—dc22

2009018200

Direct all inquiries to J. Ross Publishing, Inc., 5765 N. Andrews Way, Fort Lauderdale, FL 33309.

Phone: (954) 727-9333

Fax: (561) 892-0700

Web: www.jrosspub.com

CONTENTS

CONTRIBUTORS

Haitham M. Al-Deek
Department of Civil & Environmental Engineering
University of Central Florida

Joshua Auld
Department of Civil and Materials Engineering
University of Illinois–Chicago

Tom Brijs
Transportation Research Institute
Hasselt University

Mario Cools
Transportation Research Institute
Hasselt University

Yi Deng
Parsons Transportation Group

Emam B. Emam
URS Corporation

Jon D. Fricker
School of Civil Engineering
Purdue University

Liping Fu
Department of Civil & Environmental Engineering
University of Waterloo

Stephane Hess
Institute for Transport Studies
University of Leeds

Alan Horowitz
Department of Civil Engineering and Mechanics
University of Wisconsin–Milwaukee

Li Jin
Kittelson and Associates

Xia Jin
Cambridge Systematics, Inc.

Kara M. Kockelman
Department of Civil, Architectural, and Environmental Engineering
The University of Texas–Austin

Rongfang (Rachel) Liu
Department of Civil and Environmental Engineering
New Jersey Institute of Technology

Paul Metaxatos
Urban Transportation Center
University of Illinois–Chicago

Luis F. Miranda-Moreno
Department of Civil Engineering & Applied Mechanics
McGill University

Abolfazl Mohammadian
Department of Civil and Materials Engineering
University of Illinois–Chicago

Elke Moons
Transportation Research Institute
Hasselt University

C. Craig Morris
Bureau of Transportation Statistics
Research and Innovative Technology Administration
U.S. Department of Transportation

John W. Polak
Centre for Transport Studies
Imperial College–London

Miriam Scaglione
Institute for Economics & Tourism
HES-SO Valais University of Applied Sciences Valais–Switzerland

Filip A. M. Van den Bossche
Hogeschool-Universiteit Brussel and Katholieke Universiteit Leuven

Koen Vanhoof
Transportation Research Institute
Hasselt University

Xiaokun Wang
Department of Civil & Environmental Engineering
Bucknell University

Geert Wets
Transportation Research Institute
Hasselt University

Sadayuki Yagi
Japan Research Institute

PREFACE

In its broadest sense, the discipline of transportation statistics is the intersection of transportation and statistics. Thus, any analysis to do with transportation as well as data or statistical methods forms the study of transportation statistics. Within the purview of transportation statistics is the data that are collected and used in the statistical analysis. In fact, the role of the variety of transportation data is to determine if contemporary policies are performing as expected and identify any problematic policy issues. Such data can be used to develop the necessary policies to ameliorate the deficiencies in current transportation policy.[1]

Many advances have occurred in statistical analyses of transportation data in the past 20 years because of more efficient applications of statistical methods through advances in computing power. Because of the increased complexities in transportation phenomena, it requires the necessary data as well as statistical methods to be applied to the data. Thus, there is a complementary relationship between the statistical methods and the transportation data. This volume presents a variety of statistical methods applied to a myriad of transportation data and includes contributions that reflect international research. All of the processes are applied and illustrate how these methods can be used and interpreted. The studies should be particularly enlightening to the professional analyst in the areas of transportation research as well as transportation policy-making.

The first chapter, **Macroscopic Road Safety Modeling: A State Space Approach Applied to Three Belgian Regions** by Van den Bossche, Vanhoof, Wets, and Brijs, treats the road risk on the three main road networks in Belgium: the motorways, provincial roads, and local roads. Because of the differences in the roads—physical characteristics, number of lanes, speed limits, and others—one can expect that the road risk on these three networks will be different. The objective of this research was to analyze these data from a time series perspective, using

[1]The editor would like to thank Ashish Sen for providing valuable discussions concerning the current state of transportation statistics.

decomposition in terms of exposure and risk and using yearly data from 1973 to 2004. A multivariate state space time series model is developed in which the impact of exposure and the introduction of the major road safety laws is analyzed. The results of these models can show that road safety progress, road risk, and yearly risk reduction can be distinguished for each network.

In **Traffic Safety Study: Empirical Bayes or Hierarchical Bayes** by Miranda-Moreno and Fu, empirical Bayes (EB) methods have been widely applied for identifying hotspot locations and evaluating the effectiveness of countermeasures as a result of their sensible and uncomplicated construct in traffic studies. However, researchers advocate the use of the full Bayes (FB) approach as a better alternative to EB in such analyses. While it has been argued that FB is more flexible in accounting for full uncertainties in model parameters and accident-risk estimates than EB, little is known about their relative merits and practical differences in terms of traffic safety analyses. This chapter presents the results of a large-scale simulation study on the performance of these two approaches as applied for hotspot identification. It was found that the FB estimators performed better than the EB estimators when working with data sets with a small number of sites (observations) and characterized by an overall low mean accident frequency. However, when the data set is sufficiently large (300 or more sites), these two approaches yielded practically the same results.

Al-Deek and Emam's chapter, **Utilizing Data Warehouse to Develop Freeway Travel Time Reliability Stochastic Models,** studied four different travel time stochastic models: Weibull, Exponential, Lognormal, and Normal. The developed best-fit stochastic model (Lognormal) can be used to estimate travel time reliability of freeway corridors and report this information to the public through traffic management centers. The new method was compared to the existing methods (Florida method and Buffer Time method). Unlike the existing methods, which were insensitive to the travelers' perspective of travel time, the new method showed high sensitivity to the geographical location that reflects the level of congestion and bottlenecks. In addition, it is more appropriate for measuring the performance of freeway operations.

In transportation research, parking policy is an important component of contemporary travel-demand management policies since there is demand to park in various municipalities but there are never enough parking spaces to satisfy these demands. The chapter by Hess and Polak, **Mixed Logit Modeling of Parking-Type Choice Behavior**, explores parking policy and the effectiveness of many parking policy measures. This effectiveness depends on influencing parking-type choice, so that understanding the factors affecting these choices is of considerable practical importance. This research reports the results of an analysis of parking-choice behavior, based on a stated choice (SC) data set, collected in various city center locations in the United Kingdom. The analysis advances the state of the art in the analysis of parking-choice behavior by using a mixed multinomial logit (MMNL) model, capable of accommodating random heterogeneity in travelers' tastes and the potential correlation structure induced by repeated observations being made of the same individuals. The results of the analysis indicate that taste heterogeneity is a major factor in parking-type choice. Accommodating this heterogeneity leads to significantly different conclusions regarding the influence of substantive factors such as access, search and egress time, and the treatment of potential fines for illegal parking. It also has important effects on the implied willingness to pay for time-savings and on the distribution of this willingness across the population. Their analysis also reveals important differences in parking behavior across different journey purposes, as well as across the three locations used in the SC surveys. Finally, the chapter discusses a number of technical issues related to the specification of taste heterogeneity that are of wider significance in the application of the MMNL model.

Cools, Moons, and Wets provide an analysis of traffic counts using the effects of holidays, as this is an emerging area of research in recent years. In their chapter, **Modeling Daily Traffic Counts: Analyzing the Effects of Holidays**, two modeling approaches for forecasting daily traffic counts are compared. The starting point for the modeling is that successive traffic counts are correlated; therefore, past values provide a solid base to forecast future traffic counts. The second part of the modeling presupposes that daily traffic counts could be explained by other variables. This chapter pays special attention to the investigation of holiday effects. The analysis is performed on data collected in 2003, 2004, and 2005 from single inductive loop detectors. Results from these modeling approaches show that weekly cycles influence the variability in daily traffic counts. The Box-Tiao modeling approach provides the required framework to quantify holiday effects, with results indicating that daily traffic counts are significantly reduced during holiday periods. When the forecasting performance of the different modeling techniques was assessed, the Box-Tiao modeling approach out-performed other strategies. The results of this analysis provide opportunities for future research.

In the chapter by Metaxatos, **Issues with Small Samples in Trip Generation Estimation**, the author examines the problem facing transportation planners estimating and validating household trip generation rates from small-scale household travel surveys. Three problems are addressed: (1) unusual observations, (2) small number of observations, and (3) no observations. Unusual observations are identified using traditional methods. Classification and regression tree (CART) analysis is proposed for the second problem. Finally, the third problem is addressed using row-column decomposition analysis. The methods are demonstrated using a small-scale household travel survey and are simple enough to be implemented with the resources available to transportation analysts, especially in smaller metropolitan planning organizations.

In Chapter 7, **Recent Progress on Activity-Based Microsimulation Models of Travel Demand and Future Projects**, Mohammadian, Auld, and Yagi provide the recent developments of activity-based modeling as used in travel-demand models. There has been significant progress recently in the area of activity-based microsimulation used in travel-demand analysis. Activity-based microsimulation models have grown in importance as more criticism of traditional travel-demand models has been made. However, traditional four-step travel-demand models continue to make up the majority of the models used in practice. It is hoped that the use of activity-based models will grow with further improvements in model technique, validation, and transferability. This chapter describes some fundamental concepts of activity-based modeling and provides an overview of the various statistical procedures used in the various model components. The final part of this chapter proposes future directions for research and analyses.

Wang and Kockleman provide for an examination of several simulation methods used in a variety of empirical analyses using maximum likelihood methods. Their chapter, **Maximum Simulated Likelihood Estimation With Spatially Correlated Observations: A Comparison of Simulation Techniques**, which examines the inclusion of maximum simulated likelihood estimation (MSLE) routines in new software releases, encourages the use of simulation in estimation of complex models. It is important that analysts understand the relative performance of different simulation techniques under different data circumstances. At the same time, issues in observation correlation, such as spatial autocorrelation, are emerging in transportation (and land use) studies. Several studies have used MSLE to address spatial relationships. This chapter studies the performance of several simulation techniques with correlated observations. When a data set's true correlation patterns clearly differ from the simulated patterns, model estimation may become inefficient, and with finite samples, statistical identification of parameters may suffer. Fortunately, it is found that in most cases, even when observations are correlated, quasi-Monte Carlo (QMCs) and hybrid methods may be preferred to as pseudo-random Monte Carlo (PMC) because of their better coverage. The findings reported offer an important supplement to existing studies of MSLE simulation techniques and should prove valuable for future work that requires simulated estimation with correlated—particularly spatially correlated—observations.

Scaglione's chapter, **Analyzing the Impact of Land Transportation on Regional Tourism: The Case of the Closure of the Glion Tunnel in the Valais, Switzerland**, blends tourism and transportation. The tourism sector is important to the economy of the Valais, as some of its resorts—Zermatt, Verbier, Crans Montana, Leukerbad, and Saas Fee—are well known and draw tourism from all over the world. The Glion tunnel is two separate tunnels, one for each direction, and it was partially closed for repairs from April to November in 2004 and 2005. During each of these times, one was completely closed and traffic flowed through the other in both directions. The closings raised several questions for cantonal policy-makers and tourism planners. How many overnights did tourism lose as a result? Did tourists choose other modes of transport, such as trains? The objective of the chapter was to address these two questions using only public data like route counting, national tourism statistics, and the frequency of trains. The statistical method followed was Harvey's structural time series model. As indigenous variables, the causal version of the method uses the monthly series of tourism overnights and, as exogenous variables, used the transportation data, also with monthly frequency. To show the differences in behavior before, during, and after the closings of the tunnel, the models were recalculated with several final adjustments. Using the Chow test to identify breaking points, the approximate moments when certain changes in behavior become significant were determined.

Morris provides an application of generalized linear models along with maximum quasi-likelihood estimation to motor vehicle fatalities. In **Quasi-Likelihood Generalized Linear Regression Analysis of Fatality Risk Data**, Morris presents motor vehicle-related fatality risk as it varies in complex ways with interacting human, vehicle, and environmental factors. Statistically valid analysis of data illuminating such relationships is challenged not only by the complexity of the plausible structural models relating the fatality rate data to the explanatory variables, but also by uncertainty about the probability distribution of the data. Generalized linear models, along with maximum quasi-likelihood estimation, provide a powerful and complementary set of tools well suited to the analysis of such data. This research presents the use of generalized linear models and maximum quasi-likelihood estimation as applied in an analysis of the association of motorcyclist fatality rates with helmet laws while controlling for two climate measures—annual heating degree days and precipitation inches—as statistical proxies for residual variation in motorcyclist activity.

Realizing the benefits of statewide travel-demand models, more and more states have already been or are in the process of developing various statewide procedures, structures, or uniform models for travel-demand forecast and mode choices. The core contents of these statewide models range from land use, intercity travel, and toll facilities to freight movements and more. However, a quick scan of the existing literature reveals that little attention or activity can be found in the area of developing statewide weekend travel-demand forecasts and mode-choice models despite the widely recognized increasing travel on weekend. The research by Liu and Deng, **Developing Statewide Weekend Travel-Demand Forecast and Mode-Choice Models for New Jersey**, explores the travel patterns during the weekends. Their research intends to fill the gap by introducing an on-going effort undertaken by New Jersey Department of Transportation (NJDOT) to develop a statewide weekend travel-demand forecast and mode-choice model in New Jersey. Applying a holistic approach that balances state-of-the-art research and practical modeling applications for multiple agencies, the research team has examined the unique characteristics of weekend travel and evaluated travel survey data and is developing specifications for a statewide weekend travel-demand forecast model that can be incorporated into the existing long-range transportation planning (LRTP) process at both metropolitan and state levels.

In their chapter, **Transferability of Time-of-Day Choice Modeling for Long-Distance Trips,** Jin and Horowitz present time-of-day models that deal with the time at which travel occurs throughout the day and their research presents a study in time-of-day choice modeling for long-distance trips, with special interest in the transferability of the model. Two data sets—from the 2001 National Household Travel Survey as produced by the U.S. Department of Transportation and the 2001 California Statewide Household Travel Survey—were employed to explore the effects of various factors on time-of-day choice making and to test the transferability of the behavioral findings and model parameters. Although there are differences in data composition between the two data sets, comparative analysis of the models developed from the two data sets reveals consistent results, suggesting the potential for transferability of the behavioral pattern across spatial locations.

In the chapter, **Univariate Sensitivity and Uncertainty Analysis of Statewide Travel Demand and Land Use Models for Indiana**, Jin and Fricker apply the concepts of uncertainty as it exists in statewide travel-demand forecasting and land use models. Both input data and adopted model parameters can vary from their true values because of model misspecification, imperfect input information, and innate randomness of events. The objective of their research is to study the sensitivity of vehicles miles traveled (VMT) outputs of INTRLUDE, an integrated statewide travel-demand and land use forecasting model system for Indiana, to plausible amounts of variations on travel model parameters and input data. The land use in the Central Indiana 2 (LUCI2) model is used in conjunction with the statewide travel-demand model to predict travel demand over time through various scenarios. Results indicate that VMT outputs are the most sensitive to trip distribution function parameters and trip production rates. Population growth rates and trip assignment parameters do not have a high degree of influence on VMT outputs.

Although the discipline of transportation statistics appears to be narrow, the discipline covers a myriad of topics which encompasses a variety of data sources and statistical methods to analyze a variety of research hypotheses. This present volume provides the latest research in transportation statistics, which do not exist in a vacuum, but rather provide the study of various policies in transportation. To make transportation policy, researchers must understand the policy and know what empirical tools and data would be needed to assess this policy. Without coherent statistical methods and the appropriate data, it becomes difficult for transportation policy-makers to make informed decisions regarding a policy. The chapters in this volume give policymakers—and those who provide analysis for them—a variety of approaches of outcomes under various policy scenarios which could play a role in the development of transportation policy.

Brian W. Sloboda

ABOUT THE EDITOR

Brian W. Sloboda, Ph.D., is currently at the U.S. Postal Service as a pricing economist. He is a former economist at the U.S. Department of Transportation, Research and Innovative Technology Administration/Bureau of Transportation Statistics (BTS) in Washington D.C. While at the U.S. Department of Transportation, Brian worked in the integration of national accounts to improve the Transportation Satellite Accounts (TSAs) and the Household Production of Transportation Services (HPTS). He worked in productivity of the various modes of transportation using the Malmquist productivity index and also conducted research in the tourism sector from a time series approach. Brian was one of the committee members for the 2006 Transportation and Economic Development Conference (TED) held at Little Rock, Arkansas. Before coming to the U.S. Department of Transportation, he was an economist at the U.S. Department of Commerce, Bureau of Economic Analysis, Industry Economics Division (IED), which prepared the input-output accounts for the United States.

Additionally, he teaches at the University of Phoenix, Park University, and the University of Maryland, University College. He has also taught econometrics at the USDA Graduate School in Washington, D.C. Brian has published articles in the *Journal of Economics, Crisis Management in Tourism, Advances in Econometrics, Tourism Economics, Journal of Transportation and Statistics, Journal of Economic and Business Studies, Journal of the Transportation Forum, Journal of Business and Economic Perspectives, International Journal of Transport Economics,* and *Annals of Regional Science.*

MACROSCOPIC ROAD SAFETY MODELING: A STATE SPACE APPROACH APPLIED TO THREE BELGIAN REGIONS

Filip A. M. Van den Bossche
Hogeschool-Universiteit Brussel and Katholieke Universiteit Leuven

Koen Vanhoof
Transportation Research Institute
Hasselt University

Geert Wets
Transportation Research Institute
Hasselt University

Tom Brijs
Transportation Research Institute
Hasselt University

1.1 INTRODUCTION

Macroscopic road safety research deals with the development of road safety models to analyze the relationship between road safety and exposure at an aggregated level. In this chapter, an overview of macroscopic models is presented, and an application of state space time series modeling in road safety is given. In particular, a tri-variate state space time series model is developed to analyze the relationship between the number of victims killed or seriously injured (KSI) and the level of exposure for three regions in Belgium: Flanders, Wallonia, and Brussels-Capital. The added value of the approach is found in the fact that the victims of the three regions are analyzed together, so that the dynamics between the respective time series become clear and, at the same time, the differences in road use patterns among the three regions can be taken into account.

The use of exposure data in road safety research as a measure of the magnitude of the activity that results in fatalities is neither surprising nor new. Irrespective of the level of aggregation at which the road safety problem is studied, data on road crashes are usually analyzed together with some measure of exposure. The reasoning behind this common practice is that the number of victims counted in traffic should, in some way or another, be related to the distance traveled. When comparing road safety over different entities (road users, countries, and so on), an exposure measure is included to put the entities on the same scale. To this end, different measures of exposure can be used, depending on the per unit quantity that is required; the most frequently used measures are the number of vehicle kilometers, the population, and the vehicle park. The ratio of the number of victims or crashes, divided by the level of exposure, is usually called risk. Although it is commonly accepted that there is a relationship between the level of exposure and the level of road safety, the specific nature of this relationship is not clear. In particular, it is not straightforward to quantify the percentage of change (say β) in road safety that comes with a one percent change in exposure. In many models, this elasticity is assumed to be equal to 1, which implies a perfect proportional relationship between the number of victims and exposure, although this assumption can be easily questioned.

This chapter investigates the relationship between the number of victims and exposure, in a time series context and at an aggregated level, using yearly data. In particular, the number of victims KSI in the three Belgian regions is analyzed in a tri-variate state space model. The number of victims KSI for each region is linked to the corresponding measure of exposure—the number of vehicle kilometers driven. By including a parameter for the level of exposure, the proportionality assumption of the exposure-victims relationship can be tested and compared among the three regions. The models also include another interesting application. For each region, the trend in the number of victims KSI is considered as the combination of a trend in the road safety level and the impact of the level of exposure. This clarifies the extent to which exposure influences the number of victims. Moreover, how the level and slope components of the three regions are related is easily investigated.

The relationship between exposure, risk, and victims has been regularly investigated in road safety research, which becomes clear from the overview of macroscopic models in Section 1.2. In Section 1.3, the tri-variate state space model is presented. The data used for the model are introduced in Section 1.4. Section 1.5 describes the results of the model, whereas Section 1.6 summarizes the most important conclusions and topics for further research.

1.2 EXPOSURE AND RISK IN ROAD SAFETY RESEARCH

The concepts of exposure and risk are not new in the field of macroscopic road safety research. The modeling milestones are provided by Oppe (1989, 1991), Gaudry (1984, 2000), Lassarre (2001), and Bijleveld (2004). This section gives an overview of the various developing stages of these approaches.

1.2.1 Oppe and Its Extensions

Oppe developed a framework to analyze the relationship between exposure and risk that is based on learning theory (Oppe, 1991). He explained the evolution in fatality rates over time as the result of a collective learning process that is determined by the growth processes in exposure and risk. Exposure is seen as a growth process that is restricted in its development by the physical boundaries or maximum capacity of the road system, and is typically modeled by means of (*S*-shaped) logistic curves. For the fatality rates, a negative exponential trend is assumed, indicating a collective learning process (European Commission, 2004) caused by the ever-increasing understanding of the traffic safety problem and the constant safety performance improvement of the road transport system. The number of fatalities is then, by definition, equal to the product of the risk and exposure curves. In this line of reasoning, the relationship between exposure V_t and fatalities F_t is given by:

$$\left.\begin{aligned} \frac{F_t}{V_t} &= \exp(\alpha t+\gamma)+\varepsilon_{1,t} \\ V_t &= \frac{V_m}{1+\exp(at+c)}+\varepsilon_{2,t} \end{aligned}\right\} \Rightarrow F_t = \left(\frac{F_t}{V_t}\right)\times V_t \tag{1.1}$$

In these equations, α, γ, V_m, a, and c are parameters to be estimated, whereas $\varepsilon_{1,t}$ and $\varepsilon_{2,t}$ are the residual terms. The years are indicated by t. In the risk function, α indicates the learning rate, whereas the parameter a in the exposure function is the growth rate, and V_m is the upper boundary.

The Oppe model is an interesting starting point for the analysis of fatalities and exposure at the macro level and has been estimated for many countries (Oppe, 1991, 2001). However, the model has some shortcomings, and the classical framework has been extended by several authors. First, the Oppe model may suffer from autocorrelation in the residual terms. This was mentioned by Broughton (1991), who used a method developed by Cochrane (1949) to eliminate autocorrelation. Another approach is to estimate additional autoregressive terms simultaneously with the other parameters in the model, such as presented in Van den Bossche (2006). Second, the models are limited by the imposed functional forms; a logistic curve for exposure and an exponential curve for the risk. The logistic exposure curve reflects an upper boundary in traffic growth, but is restricted in the sense that it assumes a perfect symmetric behavior. This assumption can be relaxed by introducing more general *S*-shaped curves such as the Gompertz curve (Commandeur, 2002; Commandeur and Koornstra, 2001) or the Richards curve (Van den Bossche, 2006). Also Bijleveld and Oppe (1999) proposed some functional form extensions, allowing for more realistic relationships between fatalities and risk. A third questionable property of the Oppe model is that the exponential curve implies that the risk will eventually decrease to zero. This assumption can be relaxed by introducing a constant term in the risk function (European Commission, 2004; Van den Bossche, 2006). Fourth, some authors improved the model by introducing dummy variables to account for the introduction of major road safety measures (Cameron, 1997). Broughton (1991) proposed a model in which the introduction of compulsory seatbelt use (1983) and the impact of drink-drive legislation (1968) were tested. In Van den Bossche (2006), safety laws on seatbelt use, speed, and impaired driving were tested in an extended Oppe model for Belgium. Fifth, the Oppe model family and its extensions start from the fundamental relationship $F_t = (F_t/V_t) \times V_t$. This assumption, which inherently states that the number of victims and the level of exposure are proportional, can be relaxed by extending the relationship with a power β for V_t, which is equivalent to estimating a regression parameter for exposure that can be different from 1. The more general relationship between exposure and fatalities is then written as $F_t = r(t) \times V_t^{\beta}$, in which the component $r(t)$ is modeled with a specific functional form (typically an exponential or a logistic curve). A value of β larger than 1 implies more substantial increases in fatalities per increase in exposure compared to the proportional relationship (European Commission, 2004). The original Oppe model is a special case of this relationship, in which $r(t) = F_t / V_t$ and β is set equal to 1. Note that in this general relationship, the definition of risk, F_t / V_t would be equal to $r(t) \times V_t^{\beta-1}$, which implies that $r(t)$ in itself is not equal to the classical concept of risk. It is rather an indication of the progress that is made in road safety, apart from the impact of exposure.

1.2.2 Layered Structure of the DRAG Family

The distinctive role of exposure when modeling the number of fatalities in a time series context, together with the possibility of letting $\beta \neq 1$, is also found in the DRAG family of models (**d**emand for **r**oad use, **a**ccidents and their **g**ravity), introduced in Gaudry (1984) and Gaudry and Lassarre (2000). The DRAG models start from a layered structure in which exposure, accident frequency, and accident severity can be regressed on a set of relevant explanatory variables in an attempt to explain the development of these variables over time. DRAG models have a well-defined structure, use flexible-form regression analysis, and are calibrated with monthly time series data defined over a country or region. This (at least) three-layer recursive structure of explanation, involving road use, accident frequency and severity, is a common feature of all members of the DRAG family (European Commission, 2004).

Just as in the Oppe models, exposure is not considered as exogenous, but is modeled in a separate equation. The DRAG and Oppe models both estimate a layered structure, but contrary to the Oppe models, frequency and severity of accidents, and not the corresponding risk functions, are estimated in the DRAG models. Other differences include the explanatory power and the advanced econometric model structure of the DRAG models, including corrections for heteroskedastic and/or autocorrelated residuals. In the frequency and severity equations, a parameter for exposure can be estimated to be different from 1. Furthermore, in the DRAG model structure, Box-Cox transformations are used to relax the linearity assumption usually embedded in a regression model.

Several other authors investigated the road safety problem in explanatory models, in which exposure is one of the independent variables. For example, Christens (2003) included exposure as one of the explanatory variables in a log-linear model to explain the number of fatal accidents. However, not every explanatory model includes a parameter for exposure. Ledolter (1996) modeled the accident risk as the ratio of accident counts and the exposure level in a log-linear model with deterministic seasonal terms and a first-order autoregressive model for the noise term, thereby presuming that the coefficient of exposure is equal to 1.

1.2.3 Unobserved Components Models

All models described thus far have a deterministic curve fitting procedure in common. That is, these models assume underlying functional relationships for fatalities, exposure, and risk. Instead of imposing these deterministic curves, Lassarre (2001) introduced a stochastic local linear trend model. That is, the extended relation $F_t = r(t) \times V_t^{\beta}$ is modeled in a framework in which $r(t)$ is treated as an unobserved component that is allowed to change over time. With this approach, other patterns of evolution than an exponential or logistic one are captured

(European Commission, 2004). This stochastic trend model belongs to the family of structural models introduced by Harvey (1989). The equation for the number of victims can be written as the product of an unobserved component R_t and an exposure measure V_t, raised to a power β:

$$F_t = R_t \times V_t^{\beta} \times e_t \qquad (1.2)$$

On the log scale, this expression is a linear relationship between the log of the number of victims and the log of the level of exposure, in which an unobserved component μ_t is introduced:

$$\log(F_t) = \mu_t + \beta \log(V_t) + \varepsilon_t \qquad (1.3)$$

Using this framework of unobserved components, some authors developed models in which exposure is treated as an explanatory variable. Christens (2003) developed a state space model in which the elasticity of traffic turned out to be 1.58 percent, which is significant and larger than 1. Scuffham (2002) uses a general-to-specific approach in a state space setting on current values and lagged differences of the dependent and independent variables to model the number of fatal crashes. The current period exposure effect was highly significant and larger than 1, whereas the lagged differences had significant negative signs, indicating that, with an increased distance traveled, the risk of a crash increases at a decreasing rate. Another example of an explanatory structure for exposure is presented by Johansson (1996). He investigated the effect of a lowered speed limit on the number of accidents with fatalities, injuries, or vehicle damage. Because exposure is not known, it is modeled as a latent variable instead of using proxies such as gas deliveries or the number of registered vehicles. The latent exposure measure is determined by gasoline prices, disposable income, and an error term. The fact that exposure is a latent variable, compared to other explanatory factors, implies that the direct and indirect effects of variables are combined in one parameter, making it impossible to separate exposure and other effects. Also, the reduced form equation is identical to a model in which no exposure measure is included.

Although the models developed by Lassarre are more flexible in nature than the Oppe models, they are based on the assumption that exposure is a deterministic and faultlessly-observed variable. Therefore, another approach toward the problem is to model the level of exposure as a second dependent variable. Bijleveld and Commandeur (2004) and Gould et al. (2006) used a multivariate framework to measure accident risk and exposure simultaneously. Both risk and exposure are assumed to be unobserved, which recognizes the fact that these numbers are subject to a measurement error. Although this model formulation is in line with the conceptual framework of the Oppe models, it does not include a parameter for exposure. Therefore, the model that is presented in this chapter can be considered a multivariate extension of the model developed by Lassarre (2001).

1.3 MODEL DEVELOPMENT

The models based on the theory of unobserved components offer a high level of flexibility and are naturally linked to the basic relationship between exposure and fatalities. In the unobserved components models developed by Lassarre, a parameter β for exposure is included, which indicates a deviation from the proportional relation between exposure and the number of fatalities. This is in line with other models, among them those developed by Gaudry and Lassarre (2000), in which it was shown that the parameter estimated for exposure is not necessarily equal to 1, and should therefore be estimated in the relationship between fatalities and exposure. In this chapter, the framework proposed by Lassarre is used as a starting point, but it is extended to the multivariate case, so that the number of victims in Flanders, Wallonia, and Brussels-Capital can be investigated jointly. Based on this model, differences in trends over the three regions are revealed.

1.3.1 Tri-variate State Space Model

This section introduces the tri-variate state space model that is used to analyze the time series. Using matrix notation, a general multivariate state space model can be written as (Commandeur and Koopman, 2007):

$$\begin{aligned} y_t &= Z_t\alpha_t + \varepsilon_t, & \varepsilon_t &\sim \mathrm{NID}(0,H_t), \\ \alpha_{t+1} &= T_t\alpha_t + R_t\eta_t & \eta_t &\sim \mathrm{NID}(0,Q_t) \end{aligned} \tag{1.4}$$

The first equation is called the observation or measurement equation. Here, the 3×1 vector y_t $(t = 1,\ldots,n)$ contains the logs of the observed number of victims KSI for Flanders, Wallonia, and Brussels-Capital at time point t, and ε_t is the corresponding vector with observation errors. These errors have a zero mean and unknown variances and covariances that form the 3×3 matrix H_t. The second equation is the state equation, in which the 9×1 state vector α_t (consisting of three levels, three slopes, and three regression parameters) is updated. The matrix T_t is the transition matrix of order 9×9. The 6×1 vector γ_t contains the state errors with variances and covariances gathered in the 6×6 matrix Q_t. For the tri-variate state space model that is proposed in this chapter, the vector y_t contains the log of the $F_t^{(i)}$, which are the victims KSI for Flanders ($i = F$), Wallonia ($i = W$) and Brussels-Capital ($i = B$). The vector α_t gathers the state components, namely the levels $\mu_t^{(i)}$, the slopes $\nu_t^{(i)}$, and the parameters $\beta_t^{(i)}$, $i \in \{F, W, B\}$, which measure the exposure elasticities for the three regions.

The vectors y_t, α_t, and η_t can be written as:

$$y_t = \begin{pmatrix} \log F_t^{(F)} \\ \log F_t^{(W)} \\ \log F_t^{(B)} \end{pmatrix}, \quad \alpha_t = \begin{pmatrix} \mu_t^{(F)} \\ \mu_t^{(W)} \\ \mu_t^{(B)} \\ \nu_t^{(F)} \\ \nu_t^{(W)} \\ \nu_t^{(B)} \\ \beta_t^{(F)} \\ \beta_t^{(W)} \\ \beta_t^{(B)} \end{pmatrix}, \quad \eta_t = \begin{pmatrix} \xi_t^{(F)} \\ \xi_t^{(W)} \\ \xi_t^{(B)} \\ \varsigma_t^{(F)} \\ \varsigma_t^{(W)} \\ \varsigma_t^{(B)} \end{pmatrix} \tag{1.5}$$

whereas the matrices T_t and R_t are equal to:

$$T_t = \begin{bmatrix} 1 & 0 & 0 & 1 & 0 & 0 & 0 & 0 & 0 \\ 0 & 1 & 0 & 0 & 1 & 0 & 0 & 0 & 0 \\ 0 & 0 & 1 & 0 & 0 & 1 & 0 & 0 & 0 \\ 0 & 0 & 0 & 1 & 0 & 0 & 0 & 0 & 0 \\ 0 & 0 & 0 & 0 & 1 & 0 & 0 & 0 & 0 \\ 0 & 0 & 0 & 0 & 0 & 1 & 0 & 0 & 0 \\ 0 & 0 & 0 & 0 & 0 & 0 & 1 & 0 & 0 \\ 0 & 0 & 0 & 0 & 0 & 0 & 0 & 1 & 0 \\ 0 & 0 & 0 & 0 & 0 & 0 & 0 & 0 & 1 \end{bmatrix}, \quad R_t = \begin{bmatrix} 1 & 0 & 0 & 0 & 0 & 0 \\ 0 & 1 & 0 & 0 & 0 & 0 \\ 0 & 0 & 1 & 0 & 0 & 0 \\ 0 & 0 & 0 & 1 & 0 & 0 \\ 0 & 0 & 0 & 0 & 1 & 0 \\ 0 & 0 & 0 & 0 & 0 & 1 \\ 0 & 0 & 0 & 0 & 0 & 0 \\ 0 & 0 & 0 & 0 & 0 & 0 \\ 0 & 0 & 0 & 0 & 0 & 0 \end{bmatrix} \tag{1.6}$$

Further, the explanatory (exposure) variables $V_t^{(i)}$, $i \in \{F, W, B\}$, are found in the matrix Z_t:

$$Z_t = \begin{bmatrix} 1 & 0 & 0 & 0 & 0 & 0 & \log V_t^{(F)} & 0 & 0 \\ 0 & 1 & 0 & 0 & 0 & 0 & 0 & \log V_t^{(W)} & 0 \\ 0 & 0 & 1 & 0 & 0 & 0 & 0 & 0 & \log V_t^{(B)} \end{bmatrix} \tag{1.7}$$

and the variances and covariances of the observation equations are summarized in the matrix H_t:

$$H_t = \begin{bmatrix} \sigma^2_{\varepsilon^{(F)}} & \operatorname{cov}(\varepsilon^{(F)}, \varepsilon^{(W)}) & \operatorname{cov}(\varepsilon^{(F)}, \varepsilon^{(B)}) \\ \operatorname{cov}(\varepsilon^{(F)}, \varepsilon^{(W)}) & \sigma^2_{\varepsilon^{(W)}} & \operatorname{cov}(\varepsilon^{(W)}, \varepsilon^{(B)}) \\ \operatorname{cov}(\varepsilon^{(F)}, \varepsilon^{(B)}) & \operatorname{cov}(\varepsilon^{(W)}, \varepsilon^{(B)}) & \sigma^2_{\varepsilon^{(B)}} \end{bmatrix} \tag{1.8}$$

Letting:

$$Q_t^{(\lambda)} = \begin{bmatrix} \sigma^2_{\lambda^{(F)}} & \operatorname{cov}(\lambda^{(F)}, \lambda^{(W)}) & \operatorname{cov}(\lambda^{(F)}, \lambda^{(B)}) \\ \operatorname{cov}(\lambda^{(F)}, \lambda^{(W)}) & \sigma^2_{\lambda^{(W)}} & \operatorname{cov}(\lambda^{(W)}, \lambda^{(B)}) \\ \operatorname{cov}(\lambda^{(F)}, \lambda^{(B)}) & \operatorname{cov}(\lambda^{(W)}, \lambda^{(B)}) & \sigma^2_{\lambda^{(B)}} \end{bmatrix} \tag{1.9}$$

with $\lambda \in \{\xi, \zeta\}$ and $\mathbf{0}$ a 3×3 zero matrix, we have that:

$$Q_t = \begin{bmatrix} \mathbf{Q}_t^{(\mathrm{i})} & \mathbf{0} \\ \mathbf{0} & \mathbf{Q}_t^{(æ)} \end{bmatrix} \tag{1.10}$$

Writing out the matrices results in a set of three observation equations:

$$\begin{aligned} \log F_t^{(F)} &= \mu_t^{(F)} + \beta_t^{(F)} \log V_t^{(F)} + \varepsilon_t^{(F)}, \\ \log F_t^{(W)} &= \mu_t^{(W)} + \beta_t^{(W)} \log V_t^{(W)} + \varepsilon_t^{(W)}, \\ \log F_t^{(B)} &= \mu_t^{(B)} + \beta_t^{(B)} \log V_t^{(B)} + \varepsilon_t^{(B)} \end{aligned} \tag{1.11}$$

and nine state equations:

$$\begin{aligned} \mu_{t+1}^{(F)} &= \mu_t^{(F)} + \nu_t^{(F)} + \xi_t^{(F)}, \\ \mu_{t+1}^{(W)} &= \mu_t^{(W)} + \nu_t^{(W)} + \xi_t^{(W)}, \\ \mu_{t+1}^{(B)} &= \mu_t^{(B)} + \nu_t^{(B)} + \xi_t^{(B)}, \\ \nu_{t+1}^{(F)} &= \nu_t^{(F)} + \zeta_t^{(F)}, \\ \nu_{t+1}^{(W)} &= \nu_t^{(W)} + \zeta_t^{(W)}, \\ \nu_{t+1}^{(B)} &= \nu_t^{(B)} + \zeta_t^{(B)}, \\ \beta_{t+1}^{(F)} &= \beta_t^{(F)}, \\ \beta_{t+1}^{(W)} &= \beta_t^{(W)}, \\ \beta_{t+1}^{(B)} &= \beta_t^{(B)} \end{aligned} \tag{1.12}$$

Note that the $\beta^{(i)}$s have no associated error term in the state equations, indicating that the exposure effects are time invariant. The trend components $\mu_t^{(i)}$, $i \in \{F, W, B\}$, represent the trend in the (log) number of victims KSI for each region. Further, the estimated slopes are the growth components in the model. Since estimation is done on the log scale, the slopes can be interpreted as elasticities or percentage changes.

1.3.2 Estimation

State space models are typically handled with the Kalman filter, a method of signal processing which provides optimal estimates of the current state of a dynamic system. The basic structural state space model was presented in equation (1.4). The matrices Z_t, T_t, and R_t are assumed to be known. The state vector α_t is usually unobserved and typically contains the model parameters, such as regression coefficients, or parameters describing the state of a system (such as the level and the slope).

After formulating the model in terms of its components, the main objective usually consists of estimating the signal represented by α_t and the variances of the time-varying components. The Kalman filter can be used to estimate the unobserved vector. The Kalman recursion equations enable the calculation of the one-step forecast errors and the likelihood (Makridakis et al., 1998). This is usually accomplished in two stages (Chatfield, 2004). In the prediction stage, α_t is forecasted

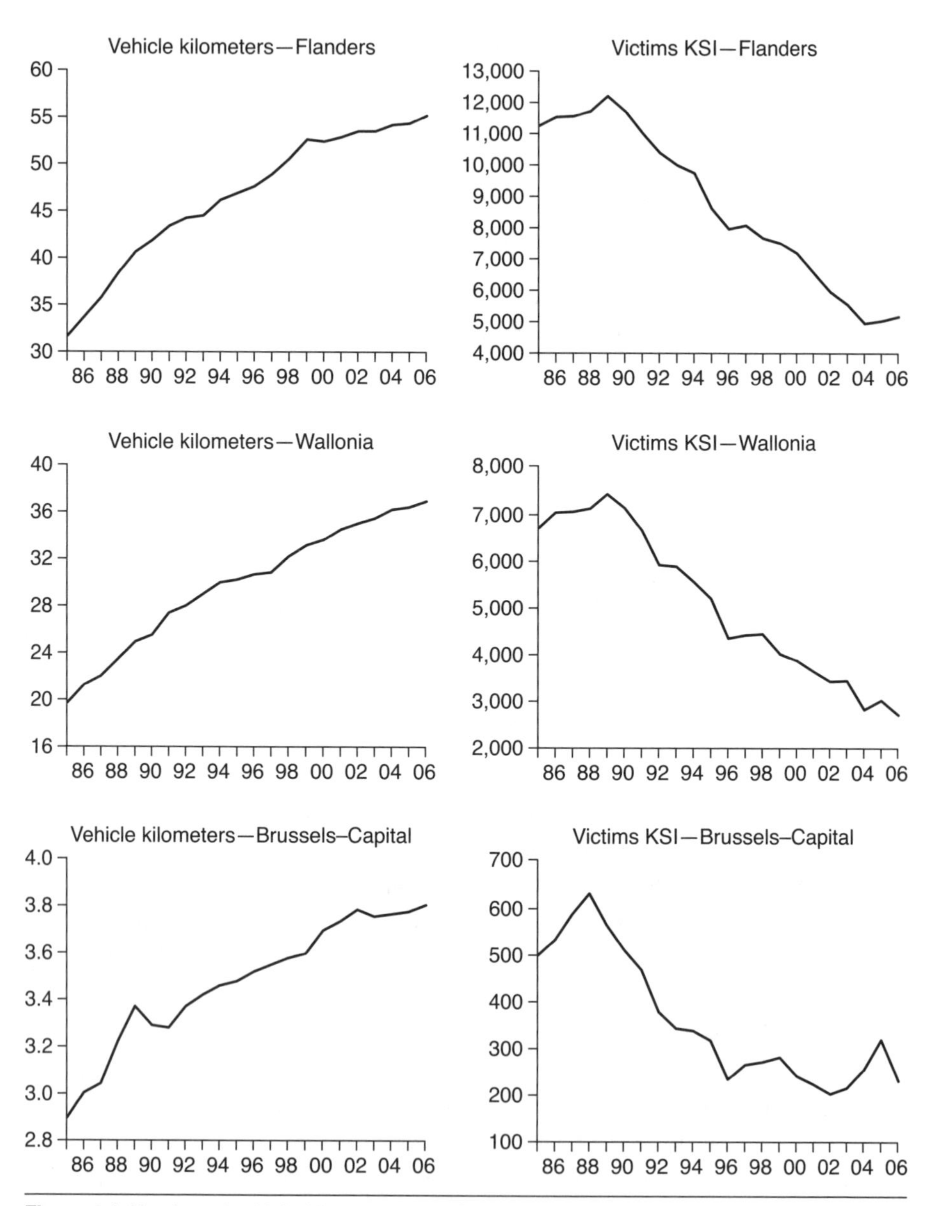

Figure 1.1 Number of vehicle kilometers ($\times 10^9$) and number of victims KSI

from the data up to time period t-1. When the new information at time t has been observed, the estimator for α_t can be modified to take into account this extra information. The prediction error of the forecast of y_t is used to update the estimate of α_t. This is the updating stage of the Kalman filter. The advantage of the recursive character of the Kalman filter is that every new estimate is based on the previous one and the latest observation, and at the same time, the entire past of the series is taken into account.

1.4 DATA

The data that are analyzed are time series of the number of vehicle kilometers (VEHKM) and the number of victims KSI in Belgium (the data are obtained from the Belgian Federal Government for Mobility and Transport and from Statistics Belgium). Three regions are considered: Flanders (F_VEHKM and F_KSI), Wallonia (W_VEHKM and W_KSI), and Brussels-Capital (B_VEHKM and B_KSI). In this study, these series are analyzed based on the assumption that the total number of victims KSI in each region is related to its own total number of vehicle kilometers. That is, the victims in Flanders are determined by vehicle kilometers driven in Flanders, the victims in Wallonia by the vehicle kilometers driven in Wallonia, and so on.

The data are shown in Figure 1.1. The graphs in the left column show the yearly number of vehicle kilometers for the three regions during 1985–2006, which increased considerably for the three regions. Comparing the values in 1985 to those in 2006, expansion factors of 76 percent for Flanders, 89 percent for Wallonia, and 32 percent for Brussels-Capital are recorded. The graphs in the right column show the numbers of victims KSI. For the Brussels-Capital region, a significant reduction in the number of victims is recorded for the last year. In Flanders, the number of serious injuries increased, whereas the number of fatalities decreased. Wallonia showed a significant decrease in serious injuries but recorded a slight increase in fatalities. Table 1.1 summarizes other characteristics of the regions (data for 2006). Flanders is the largest region and has the greatest number of inhabitants. Although Brussels-Capital is a small region, the ratio of the road length to the area and the density, defined as the number of vehicle kilometers per road length, are high. This does not come as a surprise because the capital region generates heavy traffic, especially on weekdays.

Table 1.1 Description of the three regions (2006)

	Flanders	Wallonia	Brussels
Inhabitants	6,117,440.00	3,435,879.00	1,031,215.00
Area (km^2)	13,522.00	16,844.00	161.00
Road length (km)	70,195.00	80,180.00	1,881.00
Veh. km	55.40	37.15	3.81
Road length/area	5.19	4.76	11.68
Veh. km/road length	0.79	0.46	2.03

In the models developed in this chapter, the exposure measures are linked to the number of victims KSI for the three regions. It is expected that the differences among the three regions are reflected in these models.

1.5 RESULTS

1.5.1 Model Estimation

The model presented is estimated in STAMP 6.21 (Koopman et al., 1999), using the data presented in the previous section. In the diagnostic checking phase, the one-step-ahead standardized prediction residuals of the observation equations are tested for normality and autocorrelation. To test for normality, the Doornik-Hansen (DH) statistic (Koopman et al., 1999) is used, which has a χ^2 distribution under the null hypothesis of normally distributed errors. To test for serial correlation in the residuals, the Box-Ljung Q-statistic is used at different lags. This statistic is given by (Makridakis et al., 1998):

$$Q_k = n(n+2)\sum_{k=1}^{K}\frac{r_k^2}{n-k} \tag{1.13}$$

Here, r_k is the lag-k autocorrelation coefficient, n is the number of observations, and K is the maximum lag being considered. Under the null hypothesis of no serial correlation, this test statistic is asymptotically $\chi^2_{(K-M)}$ distributed, where M equals the number of parameters estimated in the model. Large autocorrelation coefficients lead to a high Q-statistic. A high value therefore indicates significant autocorrelation and thus rejection of the null hypothesis.

In a first estimation run, it turned out that the slope disturbances were perfectly correlated. In particular, we found that $\text{corr}(\zeta_t^{(F)}, \zeta_t^{(W)}) = \text{corr}(\zeta_t^{(W)}, \zeta_t^{(B)}) = -1$ and $\text{corr}(\zeta_t^{(F)}, \zeta_t^{(B)}) = 1$. As a result, the slope of the trend component (apart from the effect of exposure) in the equations for Wallonia and Brussels can be written as a linear function of the slope of the trend component in the Flanders equation, and that the corresponding disturbances are proportional to each other (Commandeur and Koopman, 2007). Technically speaking, these findings imply that the variance matrices of the slope and level disturbances are not full rank. Similarly, $\text{corr}(\xi_t^{(W)}, \xi_t^{(B)}) = 0.9965$. Re-estimating the model using a unit rank for the slope variance matrix and a rank equal to 2 for the level variance matrix significantly improved the fit (in terms of estimated maximum of the log-likelihood function). The new model has desirable statistical properties, as can be seen in Table 1.2.

Table 1.2 Model diagnostics (normality and autocorrelation)

	DH	Q(1)	Q(6)
Flanders	0.6315 (0.7292)	0.0274 (0.8685)	10.3287 (0.1115)
Wallonia	0.4910 (0.7823)	1.1581 (0.2819)	3.9735 (0.6803)
Brussels-Capital	0.1233 (0.9402)	0.0708 (0.7902)	2.6628 (0.8498)

Note: *p*-values between parenthesis

The estimate of the variance matrix of the irregular components in the observation equations is:

$$H_t = \begin{bmatrix} 0.00033589 & 0.00071520 & -0.00009543 \\ 0.00071520 & 0.00152280 & -0.00020320 \\ -0.00009543 & -0.00020320 & 0.00002717 \end{bmatrix} \tag{1.14}$$

whereas the estimates of the components in the variance matrix for the level and slope equations are:

$$\begin{aligned} \mathbf{Q}_t^{(\xi)} &= \begin{bmatrix} 0.00072948 & -0.00012488 & -0.00022749 \\ -0.00012488 & 0.00109210 & 0.00488520 \\ -0.00022749 & 0.00488520 & 0.02200600 \end{bmatrix}, \\ \mathbf{Q}_t^{(\zeta)} &= \begin{bmatrix} 0.00013880 & -0.00000674 & 0.00005919 \\ -0.00000674 & 0.00000033 & -0.00000288 \\ 0.00005919 & -0.00000288 & 0.00002524 \end{bmatrix} \end{aligned} \tag{1.15}$$

1.5.2 Regression Parameters

Table 1.3 summarizes the regression parameter estimates, namely the parameters β_i, $i \in \{F, W, B\}$ for the exposure variables in the three observation equations. The β_i parameter estimates show the percentage of change in the number of victims KSI for a percentage of change in exposure. It is interesting to note that the victims KSI data for the three regions react differently to changes in exposure. In Flanders, the parameter is significant and much larger than 1, indicating a more than proportional (1.82 percent) increase in victims KSI for a 1 percent increase in exposure. For Wallonia, a highly significant elasticity of 1.34 is found. The exposure parameter for the Brussels-Capital region, however, has an opposite outcome and is not significantly different from zero.

Table 1.3 Exposure regression parameter estimates

	Coefficient	Root mean square error (RMSE)	*t*-value	Probability
Flanders	1.8151	0.4584	3.9596	0.0008
Wallonia	1.3382	0.2433	5.5001	0.0000
Brussels-Capital	−0.9460	1.4845	−0.6373	0.5312

The insignificant result for the Brussels-Capital region can be explained in several ways. First, the number of fatalities in the Brussels-Capital region is not influenced by exposure, which implies that an increase or decrease in the level of exposure does not result in a change in the number of victims. This is, however, unlikely. It is known that the level of exposure is a determining factor in the road safety level of a region, especially in Brussels-Capital where the level of exposure is high compared to the available road length. Second, the elasticity, which is the relationship between the log of the number of victims KSI and the log of exposure, is significant but not constant. The constant elasticity assumption was entered into the model through the log-log relationship and can be questioned. Especially for the Brussels-Capital region, one can expect that the number of victims KSI first increases with exposure, but decreases after a certain point of saturation of the network. The Brussels-Capital region is characterized by daily congestion problems, which reduce speed and hence the number of victims. A kind of inverse U-shaped relationship between the number of victims KSI and the level of exposure is, then, more appropriate. The inverse U-shaped relationship can be tested by entering a quadratic term into the model, although this implies a perfectly symmetric shape, which is again questionable. An alternative is to make β a time-varying parameter. This does not impose a strict functional form and indicates how the relation deviates from linearity. Although an extension of the model might refine the estimates, the fact that we do not find an elasticity larger than 1 for the Brussels-Capital region is in line with our expectations.

1.5.3 Model Components

An advantage of the state space approach to time series modeling is that the various components of the series are explicitly modeled. Whereas in traditional autoregressive integrated moving average (ARIMA) modeling, a trend is filtered away to render the series stationary, it is included in a state space model as a separate component that is allowed to change over time. The trend components can be transformed into an index to compare the road safety progress for the three regions, under the assumption of constant elasticity and apart from the impact of exposure (see Figure 1.2). All trends are rescaled to start at 1 in 1985. The curves,

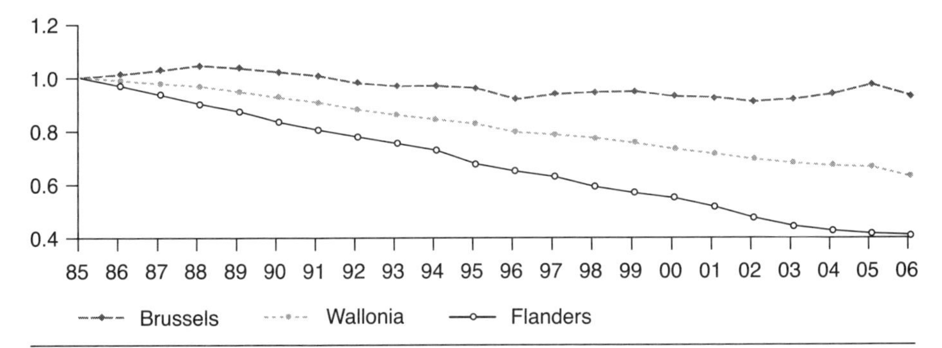

Figure 1.2 Index for the trend components

therefore, show how road safety in the three regions evolves relative to each other. Clearly, the progress for Flanders is greater than for Wallonia and Brussels. Less progress in road safety, together with relatively large increases in exposure, leads to increases in the number of victims.

Another output from the model is the slope component for the three modes of transport. Because of the high correlation between the disturbances of the slope components, the rank of the slope variance matrix was reduced to 1, and the following relationships were found:

$$\begin{aligned} \nu_t^{(W)} &= -0.0486\nu_t^{(F)} - 0.0895, \\ \nu_t^{(B)} &= 0.4265\nu_t^{(F)} + 0.0125 \end{aligned} \tag{1.16}$$

These relationships show that the road safety growth components (apart from the exposure effect and under constant elasticity assumption) for the three regions are strongly- and linearly-related (see Figure 1.3). The slope of the Wallonia trend is read from the right axis, the slope of the Brussels trend is read from the left axis. When one moves along the X-axis, the road safety improvement for Flanders decreases (that is, becomes less negative). At the same time, the road safety improvement increases (slightly) in Wallonia and decreases in Brussels. This shows that the road safety improvements in the three Belgian regions do not coincide. A stronger growth in one region can coincide with a weak growth in another region.

Two notes should be made here. First, the relationship between the slopes does not imply that a growth in one region corresponds to a decrease in another. It is perfectly possible (and frequently the case) that all growth rates have the same sign. But the differences in magnitude are related. Second, these relationships do not include the exposure effect because the slope shows only the change per time unit in the trend component. The relationship between the slopes of the different

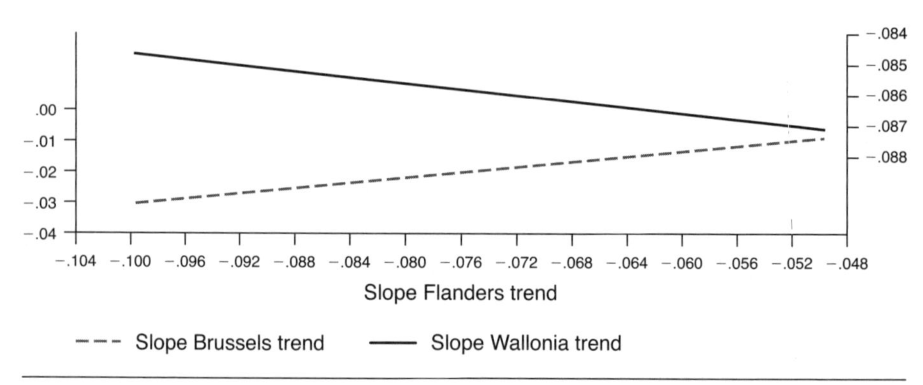

Figure 1.3 Relation between slope components

modes of transport certainly deserves further attention, especially when a country-wide road safety policy is monitored.

1.6 CONCLUSIONS

In this chapter, a tri-variate state space model was developed to analyze the relationship between the number of victims KSI and the level of exposure for three regions in Belgium: Flanders, Wallonia, and Brussels-Capital. The model fits into the stream of international research on the topic of macroscopic road safety models. The added value of this approach is found in the fact that various types of victims are analyzed together, so that the dynamics between the series become clear. Moreover, the level and slope components in the model are treated stochastically, which implies that they are allowed to vary over time.

For each region, a parameter is estimated to quantify the percentage change in the number of victims as a result of a percentage change in exposure. The model shows that the effect is different for the three regions. In particular, it is high for the Flanders and Wallonia regions and not significant for the Brussels-Capital region. However, the latter result does not necessarily imply that there is no relationship. It is, rather, an indication of the fact that the constant elasticity assumption might not be correct. The results also indicate the importance of estimating the exposure parameter, allowing for nonproportional relationships between exposure and the number of victims KSI. It should be noted that the remaining component in the model, in this case an unobserved trend component, does not correspond to the classical concept of risk (the number of victims KSI divided by the level of exposure). Indeed, when the coefficient is different than 1, the risk level is influenced by the level of exposure. The classical risk definition is only obtained when $\beta = 1$, which is obviously not the case. This conclusion can have important consequences for the development of macroscopic road safety models and should be investigated further in a more general context.

The fact that the slopes of the trends for the three transport modes are perfectly correlated is another interesting output from the model. The consequence is that the growth level for the Wallonia and Brussels region can be predicted from the growth level for the Flanders region (apart from the exposure effect). This result has important policy implications. For a given reduction in the number of victims in Flanders, the reduction in the Wallonia and Brussels victims KSI is determined by the relationship between the slope components. This implies that the slope restrictions offer benchmarks on the growth rates among regions.

The presented approach toward the relationship between exposure and the number of victims enhances the practical aspect of macroscopic road safety modeling. Therefore, it is worthwhile to investigate these issues further. From the modeling point of view, the tri-variate model can be extended to a framework in which the exposure effect is treated as a stochastic component, which challenges the constant elasticity assumption. From a road safety point of view, the relationship between the given model and the classical concept of risk needs further investigation, including the way in which exposure can influence the level of risk. Also, the practical usefulness of the model can be explored further by testing other cohorts of victims, or by comparing various countries within this framework.

REFERENCES

Bijleveld, F. and J. Commandeur. 2004. The basic evaluation model. In ICTSA meeting, INRETS, Arceuil, France.

Bijleveld, F. and S. Oppe. 1999. Aggregated models for traffic safety. In COST 329 Reader. European Commission, Directorate General for Transport.

Broughton, J. 1991. Forecasting road accident casualties in Great Britain. *Accident Analysis and Prevention*, 23(5): 353–362.

Cameron, M. 1997. Review of modelling of road casualties in the Netherlands. *Technical Report D-97-18*, SWOV.

Chatfield, C. 2004. *The Analysis of Time Series—An Introduction*, 6th ed. Boca Raton, FL: Chapman & Hall/CRC.

Christens, P. F. 2003. "Statistical modelling of traffic safety development." PhD thesis, Lyngby, Denmark: Technical University of Denmark and Danish Transport Research Center.

Cochrane, D. and G. Orcutt. 1949. Application of least squares regression to relationships containing autocorrelated error terms. *Journal of American Statistical Association*, 44: 32–61.

Commandeur, J. 2002. Algemene en periodieke trends in de ontwikkeling van de verkeersveiligheid in acht ontwikkelde landen. *Technical Report R-2002-17*, SWOV.

Commandeur, J. and S. Koopman. 2007. *An Introduction to State Space Time Series Analysis. Practical Econometrics*. Oxford University Press.

Commandeur, J. and M. Koornstra. 2001. Prognoses voor de verkeersveiligheid in 2010. *Technical Report R-2001-09*, SWOV.

European Commission. 2004. Cost 329: Models for traffic and safety development and interventions. Technical report, Directorate General for Transport.

Gaudry, M. 1984. Drag, model of the demand for road use, accidents and their severity, applied in Quebec from 1956 to 1982. *Technical Report AJD-17*, Universit de Montral–Agora Jules Dupuit.

Gaudry, M. and S. Lassarre. 2000. *Structural Road Accident Models: the International DRAG Family*. Oxford, UK: Elsevier Science.

Gould, P., F. Bijleveld, J. Commandeur, and S. Koopman. 2006. A latent variable model of traffic accident risk. In ICTSA Meeting, SWOV, Leidschendam.

Harvey, A. 1989. *Forecasting, Structural Time Series Models and the Kalman Filter*. Cambridge, UK: Cambridge University Press.

Johansson, P. 1996. Speed limitation and motorway casualties: a time series count data regression approach. *Accident Analysis and Prevention*, 28(1): 73–87.

Koopman, S., A. Harvey, J. Doornik, and N. Shephard. 1999. *Structural Time Series Analysis, Modelling and Prediction Using STAMP*, 5th ed. London, UK: Timberlake Consultants Press.

Lassarre, S. 2001. Analysis of progress in road safety in ten European countries. *Accident Analysis and Prevention*, 33: 743–751.

Ledolter, J. and K. Chan. 1996. Evaluating the impact of the 65 mph maximum speed limit on Iowa rural interstates. *The American Statistician*, 50(1): 79–85.

Makridakis, S., S. Wheelwright, and R. Hyndman. 1998. *Forecasting: Methods and Applications*, 3d ed. John Wiley & Sons.

Oppe, S. 1989. Macroscopic models for traffic and traffic safety. *Accident Analysis and Prevention*, 21: 225–232.

———1991. The development of traffic and traffic safety in six developed countries. *Accident Analysis and Prevention*, 23(5): 401–412.

———2001. Traffic safety developments in Poland. *Technical Report D-2001-8*, Stichting Wetenschappelijk Onderzoek Verkeersveiligheid SWOV.

Scuffham, P. 2003. Are there seasonal unit roots in traffic crash data? *International Journal of Transport Economics*, 30(3): 355–362.

Scuffham, P. and J. D. Langley. 2002. A model of traffic crashes in New Zealand. *Accident Analysis and Prevention* 34(5): 673–687.

Van den Bossche, F. 2006. "Road safety, risk and exposure in Belgium: an econometric approach." PhD thesis, Hasselt University.

2

TRAFFIC SAFETY STUDY: EMPIRICAL BAYES OR HIERARCHICAL BAYES?

Luis F. Miranda-Moreno
Department of Civil Engineering & Applied Mechanics
McGill University

Liping Fu
Department of Civil & Environmental Engineering
University of Waterloo

2.1 INTRODUCTION

In traffic safety studies, empirical Bayes (EB) methods have been widely applied for identifying hotspot locations and evaluating the effectiveness of countermeasures due to their rational and uncomplicated construct. Recently, however, some traffic safety researchers have been advocating the hierarchical (full) Bayes (HB) approach as a better alternative to EB for analyses of accident data. Whereas it has been argued that HB is more flexible in accounting for all uncertainties in model parameters and different model settings than EB, little is known about their relative merits and practical differences in terms of traffic safety analyses. This chapter presents the results of a simulation study on the performance of these two approaches as applied to hotspot identification. It was found that the HB approach performed slightly better than the EB approach when working with data sets containing a small number of sites (for example, a sample of 50 locations) and characterized by an overall low mean-accident frequency. However, when the data set is sufficiently large (for example, over 300 sites), these two approaches yielded practically the same results.

In traffic safety literature, Bayesian methods are commonly adopted for tasks such as ranking and selection of hotspots for engineering safety improvements and evaluation of countermeasures in before and after studies such as Hauer (1997), Schluter et al. (1997), Higle and Witkowski (1988), Persaud et al. (1999), Heydecker and Wu, (2001), Tunaru (2002), Miaou and Song (2005), Miranda-Moreno (2005), Song et al. (2006), and Brijs et al. (2007). Within the class of Bayesian methods applied in traffic safety analysis, we can distinguish two main approaches, empirical Bayes (EB) and hierarchical Bayes (HB)—also referred to as the full Bayes approach in the traffic safety literature. One important difference between these two approaches is in the way the model parameters are determined.

In the EB approach, the parameters are estimated using a maximum likelihood technique or any other techniques (for example, method of moments), involving the use of the accident data. In this approach, the estimates depend on the data only. The EB approach in road safety studies is commonly implemented using negative binomial (NB) models with a fixed or with a varying dispersion parameter as a function of some site-specific attributes (for instance, see Miranda-Moreno et al., 2005). The popularity of the NB models is due to its computational advantages because its posterior distribution has a close form (Gamma) with easier posterior inference than its competitors (such as the Poisson/Lognormal model). After its introduction to the traffic safety field, the EB approach was quickly adopted because it accounts for the non-linear relationship between the number of accidents and traffic volumes, it corrects for regression to the mean (RTM),[1] and it has a simple mathematical form and, therefore, is relatively easy to apply (Hauer, 1997).

In a HB approach, the parameters of the prior distributions (also called hyper-parameters) are fixed by modelers. For example, hyper-parameters can be specified without using any prior information or past experiences with the behavior of the data. In the traffic safety literature, the HB approach is normally implemented through hierarchical Poisson Bayes models. This approach has received increased interest in traffic safety research during the past few years due to its modeling flexibility. Its increasing popularity is also attributed to increasing desktop computing power and the availability of flexible software packages (for example, WinBUGS) for the Bayesian analysis of complex statistical models (Carlin and Louis, 2000; Gelman et al., 2003).

Despite its popularity, the EB approach has been criticized for its inability to incorporate uncertainties in the model parameters and its lack of justification for using the data twice.[2] This approach assumes the parameters are error free and can be replaced simply by their estimates for the posterior analysis. On the other hand, the HB approach overcomes this limitation with a more flexible modeling framework, but it requires the analysts to be involved directly in the model calibration process using some advanced statistical software packages (see Gelman et al., 2003; Rao, 2003). The obvious consideration is whether it is worth adopting the HB approach, which is more powerful but also more time consuming, as an alternative to the commonly used EB approach for safety studies. Under what conditions do the differences between these methods become negligible? Few studies in the traffic safety literature have investigated the practical implications of using these alternative approaches in the final output of the analysis in terms of hotspot detection capability and precision of parameter estimates.

The objective of this chapter is twofold: (1) to provide an overview of accident frequency models and Bayesian methods in the traffic safety context and (2) to conduct a simulation study for evaluating the impact of the use of these two alternative Bayesian approaches under different crash-data scenarios. This simulation study is designed to generate data that can be used to evaluate the performance of EB estimators derived from the traditional negative binomial (NB) model and HB estimators from a hierarchical Poisson/Gamma model. In doing so, a comparison is made on estimated versus true parameters and ranks.

2.2 BAYESIAN ANALYSIS FOR SAFETY STUDIES

To introduce the Bayesian analysis, we define Y_i as the number of accidents at site i (i = 1,..., n sites) over a given period of time. It is assumed that Y_i follows a Poisson distribution with a density function $f(.|\theta_i)$, where θ_i denotes the mean accident frequency.[3] Our intent is to make inferences about θ_i based on the crash input data. In a Bayesian framework, a prior distribution on the parameter $\grave{e}_i$ is first assumed, denoted as $\partial\ (\grave{e}_i|\eta)$, where η is a vector of prior parameters. This prior information is then combined with the information brought by the sample: the observed number of accidents, denoted by y_i, into the posterior distribution, represented by $p(\grave{e}_i|y_i, \eta)$. The posterior distribution of θ_i is a direct result of Bayes' theorem and has the following form:

$$p(\theta_i \mid y_i, \eta) = \frac{f(y_i \mid \theta_i)\pi(\theta_i \mid \eta)}{m(y_i \mid \eta)} \tag{2.1}$$

(Carlin and Louis, 2000)

where $f(y_i|\theta_i)$ is the likelihood of the accident frequency data, and the quantity $m(y_i|\eta)$ is referred to as the marginal distribution of y_i. This is computed by integrating the probability density function of observing y_i over the parameter θ_i.

Generally, two steps are necessary to perform a Bayesian analysis. The first is to specify a prior distribution on the mean number of accidents, θ_i. The second is to perform the usual Bayesian update with the prior distribution and historical accidents at a site to derive the posterior distribution (Carlin and Louis, 2000; Rao, 2003). The choice of prior distribution can be a challenge in practice, and several methods are available (Berger, 1985). However, a common type of priors used in road safety and many other areas is the conjugate priors. These are chosen frequently because the posterior and prior distributions have the same form. For example, in the common Poisson/Gamma model, the posterior distribution of θ_i is also distributed according to a Gamma, leading to a so-called conjugate model.

2.3 CRASH FREQUENCY MODELING

For crash-frequency analysis, the use of count data models is a common practice. In addition, to account for unobserved (unmeasured) heterogeneity and to address the over-dispersion problem, the introduction of random variations is usually considered. This leads to the random effect models such as the standard negative binomial model and its extensions, as detailed in the following section.

2.3.1 Negative Binomial Model

Assuming again that the number of accidents Y_i at site i over a given time period $T_{i,}$ is Poisson distributed, the standard NB or Poisson/Gamma model for Y_i can be represented as:

$$\begin{aligned} \text{i. } & Y_i | T_i \cdot \theta_i \sim Poisson(T_i \cdot \theta_i), \\ & \qquad \sim Poisson(T_i \cdot \mu_i\, e^{\varepsilon_i}) \\ \text{ii. } & e^{\varepsilon_i} | \phi \sim Gamma\,(\phi, \phi) \end{aligned} \tag{2.2}$$

(Winkelmann, 2003)

where $e^{\delta j}$ represents the multiplicative random effect of the model and μ_i is commonly specified as a function of site-specific attributes or covariates as follows:

$$\mu_i = f(F_{i1}, F_{i2}\ \boldsymbol{x}_i; \boldsymbol{\beta}) \tag{2.3}$$

where, F_{i1} and F_{i2} are traffic flows representing a measure of traffic exposure[4], $\boldsymbol{x}_i$ is a vector of covariates representing site-specific attributes and $\boldsymbol{\beta} = (\beta_0, \ldots, \beta_k$ is a vector of regression parameters estimated from the data. To capture the nonlinear relationship between the mean number of accidents and traffic flows, alternative functional forms can be specified for $\grave{\imath}_i$. For instance, a common functional form for intersections is:

$$\mu_i = \beta_0 (F_{1i} + F_{2i})^{\beta_1} (F_{2i} / F_{1i})^{\beta_2} \exp(\beta_3 x_{i3} + \ldots + \beta_k x_{ik}) \tag{2.4}$$

(Miaou and Lord, 2003)

This functional form follows the logic of *no traffic flows, no accidents*, permitting the relationship between accidents and traffic flows to be nonlinear, which is advocated by most studies. A variety of functional forms have been investigated extensively in the past, beginning in the 1950s. (For a detailed discussion on functional forms, refer to Miaou and Lord, 2003). When modeling accident data from road sections, the road segment length (L_i) can be incorporated as an offset (Qin et al., 2004). In addition, the two-way average annual daily traffic (AADT) is usually observed representing the sum of F_{1i} and F_{2i}. Other functional forms for intersections and non-intersection sites are discussed in Miaou and Lord (2003) and Qin et al. (2004).

The term $Gamma(\gamma)$ in Equation (2.2) denotes a Gamma probability density function[5] with parameters $\phi > 0$. More details of the Gamma distribution are presented in Cameron and Trivedi (1998). This distribution ensures that $\theta_i > 0$ because $e^{\varepsilon_i} > 0$. The parameter ϕ is usually defined as the inverse dispersion parameter, and $\alpha = 1/\phi$ is usually defined as the dispersion parameter or over-dispersion parameter. Finally, T_i is the time period of observation, which can be considered as fixed if it is constant across locations.

After the integration of the random error e^{ε_i} or θ_i, the marginal distribution of Y_i appears to be the density of the NB distribution. The log-likelihood of the NB model can then be maximized numerically using the Newton-Raphson algorithm. Some of the applications of this model in the context of hotspot identification were presented by Abbess (1981), Hauer and Persaud (1987), and Higle and Witkowski (1988).

2.3.2 Extensions of the Negative Binomial Model

Although the NB regression model is the most popular for modeling accident data, a simple extension of this model allows variability in the dispersion parameter α. Some recent work on traffic safety modeling has shown that the variance structure can depend on some covariates (Heydecker and Wu, 2001; Miaou and Lord, 2003; Miranda-Moreno et al., 2005). Such a model is called a heterogeneous NB or a generalized negative binomial model (GNB), allowing the over-dispersion parameter (α as α_i) to differ by site. In particular, α_i can be a function of a vector of site attributes ($\boldsymbol{w}_i = w_{i1}, ..., w_{im}$) that may or may not include covariates of the vector $\boldsymbol{x}_i$. The vector $\boldsymbol{w}_i$ determines the magnitude of α_i, which can be modeled using an exponential link function, so that, $\alpha_i = \exp(\boldsymbol{w}_i\boldsymbol{\gamma})$, where $\boldsymbol{\gamma} = (\gamma_0, ..., \gamma_m)$ is a vector of parameters. As in the NB model, the expectation and variance of the marginal distribution of this model is given by $E(y_i| \mu_i, \alpha_i) = \mu_i$ and $\text{Var}(y_i| \mu_i, \alpha_i) = \mu_i + \alpha_i\mu_i^2$, respectively. This model assumption allows more flexibility to deal with over-dispersion problems than its competitor. Other empirical studies have used alternative count data models such as the zero-inflated-negative binomial,

Gamma, and Poisson-underreporting models to handle problems such as excess of zeros, underdispersion, and underreporting problems, respectively.

2.3.3 Hierarchical Poisson Models

Considering again that Y_i, is Poisson distributed, a hierarchical Poisson/Gamma model can be defined as follows:

$$\begin{aligned} &\text{i. } Y_i|T_i,\theta_i \sim Poisson(T_i \cdot \theta_i), \\ &\qquad\qquad \sim Poisson(T_i \cdot \mu_i\, e^{\varepsilon_i}) \\ &\text{ii. } e^{\varepsilon_i}|\phi \sim Gamma\,(\phi, \phi) \\ &\text{iii. } \phi \sim Gamma(a, b) \end{aligned} \tag{2.5}$$

(Rao, 2003)

where μ_i is defined by Equation (2.3), that is, $\mu_i = f(F_{i1}, F_{i2}, \boldsymbol{x}_i; \boldsymbol{\beta})$; e^{ε_i} follows a Gamma distribution with $E[e^{\varepsilon_i}] = 1.0$ and $\text{Var}[e^{\varepsilon_i}] = 1/\phi$ and ϕ is also assumed to be random following a Gamma distribution with a and b as its parameters. Furthermore, the regression coefficients, $\boldsymbol{\beta}$, are also assumed to be random. As a result, this third level of randomness assumed on model parameters, ϕ and $\boldsymbol{\beta}$, is one of the main differences between this hierarchical model and the NB model discussed in the previous section. An informative or noninformative prior can be assumed on ϕ. For instance, an informative prior can be assumed by fixing the hyperparameters a and b, according to data collected from past studies or expert knowledge (Miranda-Moreno et al., 2007). In addition, the regression parameters are commonly assumed to be noninformative. Note that instead of assuming a Gamma distribution as a prior for e^{ε_i}, a hierarchical Poisson/Normal model is derived by assuming a Normal distribution on ε_i, that is, $\varepsilon_i = \log(e^{\varepsilon_i} \mid \sigma^2 \sim N(0, \sigma^2)$, with a proper hyper-prior for the parameter σ^2 (Rao, 2003). An alternative formulation is the Poisson/Lognormal with a Lognormal prior in the multiplicative random error, $e^{\varepsilon_i} \sim \text{Log}N(-0.5\tau^2, \tau^2)$ with a proper hyper-prior for the parameter τ^2 (see Winkelmann, 2003; Miranda-Moreno, 2006).

To obtain parameter estimates, posterior inference can be completed by Markov Chain Monte Carlo (MCMC) simulation methods such as Gibbs sampling and Metropolis-Hastings algorithm (see Gelman et al., 2003), which are already implemented in the software package WinBUGS.

2.3.4 Extensions of Hierarchical Models

One advantage of the hierarchical Poisson models is that they can consider spatio-temporal patterns of accident counts. This model extension is useful for working with more complex data sets in which longitudinal and/or spatial information is available (Miaou and Song, 2005; Song et al., 2006). Space-time hierarchical Poisson models can be defined by incorporating into the mean number of accidents a spatial and temporal effect, for example, $\theta_{it} = \mu_i e^{\varepsilon_{it}} e^{\varphi_t} e^{\zeta_i}$, where θ_{it} is the mean number of accidents at site i and period t (for each site there are t observation periods $t = 1,\ldots, T$). In this model, $\varphi_t(t = 1,\ldots, T)$ denotes a time effect representing a possible time trend due to socioeconomic changes, traffic operation modifications, weather variations, and so on, ζ_i $(i = 1,\ldots n)$ is a space-random effect accounting for spatial correlations among sites. This effect attempts to model spatial dependency of θ_{it} with respect to other sites, which suggests that sites that are closer to each other are likely to have common features affecting their accident occurrence. The spatial variations contained in ζ_i are not captured by the available covariates or any model component. As noted by Miaou and Song (2005), random variations across sites can be structured spatially in some way due to the complexity of the traffic interaction around locations or given that driver behaviors are influenced by a multitude of factors (for example, roadside development near locations and environmental conditions). Hierarchical Poisson models have also been extended to a multivariate setting that has been used to analyze several accident types simultaneously (for example, accident counts divided in different severity levels such as fatal accidents, accidents with injuries, among others). Under a multivariate setting, one is able to take into account correlation structures between the response variables. For instance, we suspect that a given site with a high number of fatal accidents also presents a high number of injury accidents because they can share the same set of risk factors (for example, an excessive curvature or poor road surface condition). Multivariate Poisson hierarchical models with covariates and time-space random effects have been studied by Song et al. (2006). Without accounting for spatial correlation, Tunaru (2002) and Brijs et al. (2007) have also analyzed accident data sets with different severity outcomes using a multivariate Poisson setting. Additionally, latent class (mixture) hierarchical models have been explored by Miranda-Moreno (2006) and, more recently, Lord et al. (2008b) have evaluated the application of the Conway-Maxwell-Poisson model in a HB setting.

2.4 POSTERIOR ANALYSIS

As discussed previously, one of the main objectives of a Bayesian analysis is to estimate the safety status of locations of interest under particular conditions. These estimates are the basis for safety analysis activities, including identifying dangerous locations (or hotspots) and evaluating the effectiveness of countermeasures in before-after analysis. As a result, the performance of alternative Bayesian approaches should be examined in terms of their effects on the final outcome of these safety analysis activities. In this research, we focus on one of these activities: hotspot identification.

Various ranking criteria or safety measures can be used for hotspot identification. Among the most popular are the posterior expected number of accidents, the posterior probability of excess, the posterior expectation of ranks, and the posterior probability that one site is the worst. These safety measures have been used in previous works (Schluter et al., 1997; Heydecker and Wu, 2001; Tunaru, 2002).

The posterior expectation of θ_i, denoted here by $\widetilde{\theta}_i$, is perhaps the most popular in the safety literature and is defined as:

$$\widetilde{\theta}_i = E[\theta_i \mid y_i] = \int_0^{\infty} \theta_i p(\theta_i \mid y) d\theta_i \tag{2.6}$$

This ranking criterion is a point estimate of the underlying mean number of accidents over the long term. For instance, if $\widetilde{\theta}_i$ is equal to two accidents per year, it means that in 5 years, 10 accidents are expected. To select a list of sites for safety inspections, we can simply sort the n sites under analysis based on their posterior mean number of accidents and then select the top r locations ($r < n$) as hotspots. When using an EB approach, we can compute $\widetilde{\theta}_i$ as:

$$\widetilde{\theta}_i = w_i \mu_i + (1 - w_i) y_i, \text{ where } w_i = 1/(1 + \mu_i / \varphi_i) \tag{2.7}$$

(Hauer, 1997)

An alternative accident risk measure is the posterior of excess defined as $p(\theta_i \geq c \mid y_i)$ where c stands for a critical or standard value such as the observed mean number of accidents in the locations under analysis. Another more complex measure is the posterior distribution of ranks or posterior probability of selecting the worst site (see, for instance, Tunaru, 2002; Miaou and Song, 2006). This criterion can be computed easily in a hierarchical Bayes setting when using hierarchical Poisson models.

2.5 SIMULATION STUDY

A simulation-based framework is built to compare the performance of the EB and HB approaches introduced in the previous section. The advantage of working with simulated crash data is that we know the true safety state and true ranks of each location. As a result, we know if each location is a hotspot (for example, Cheng and Washington, 2005). Hence, the true ranks are obtained from the true accident frequency distribution, which are then compared with the estimated ranks determined according to an EB or hierarchical Bayes method. The performance of alternative accident risk methods is then evaluated by making a comparison between the estimated versus true ranks. That is, a comparison is made according to the EB and HB ranks versus the ranks obtained with the true model. Finally, the ranking differences can be compared to evaluate the performance of each estimator using some measures such as Spearman correlation coefficient and percentage deviation as defined above.

Despite the fact that different models and posterior measures can be used for the hotspot identification analysis, our simulation study considers the most popular model and ranking criterion: the Poisson/Gamma model and the posterior mean of accident frequency.

2.5.1 Simulation Design

The simulation procedure designed for assessing the performance of the two alternative Bayesian estimators includes the following steps (see Figure 2.1):

Step 1. Select a random set of n sites (for example, $n = 100$) from an existing inventory database with known site-specific attributes ($\boldsymbol{x}_i$) for each site i.

Step 2. Assume that accident occurrence at the selected locations follow a Poisson/Gamma model with known parameters ϕ and $\boldsymbol{\beta}$. Based on these

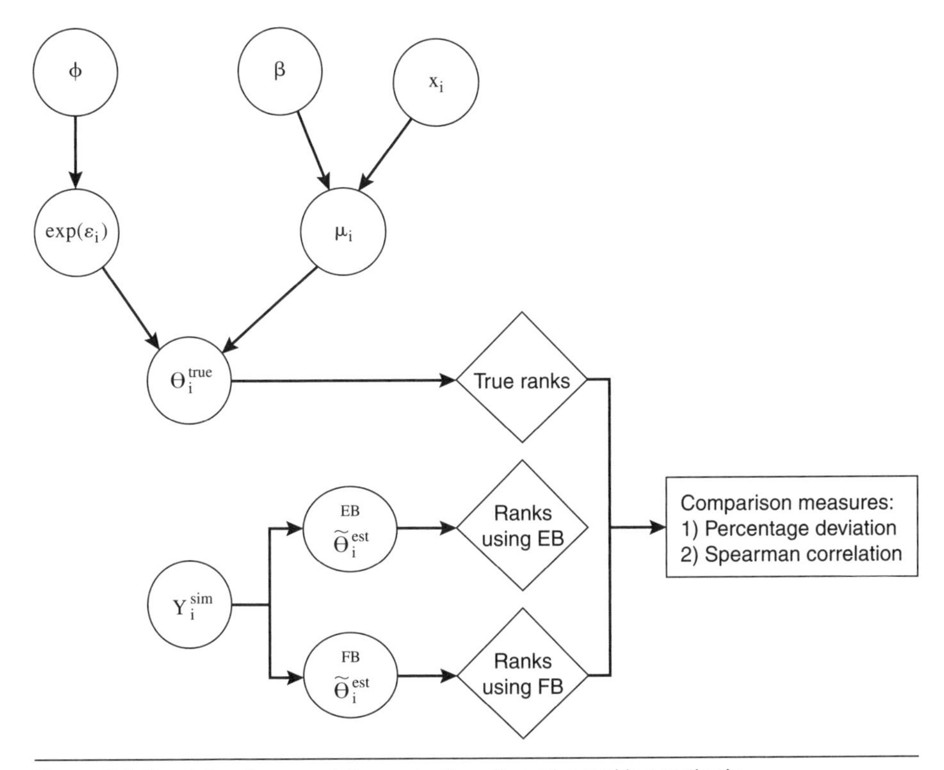

Figure 2.1 Simulation framework for evaluating alternative ranking methods

parameters and the site-specific attributes $\boldsymbol{x}_i$, generate the true mean accident frequency for each site i as follows:

$$e^{\varepsilon_i} \mid \phi \sim Gamma(\phi,\phi),$$
$$\theta_i^{true} = \mu_i e^{\varepsilon_i} \text{ for } \mu_i = f(x_i';\beta)$$

Step 3. Sort the $\boldsymbol{\theta}^{true}$ vector and obtain the true ranks:

$$\text{True mean: } \{\theta_{[1]}^{true} > \theta_{[2]}^{true} > \ldots > \theta_{[n]}^{true}\}$$
$$\text{True ranks: } \{[1], [2], \ldots, [n]\}$$

As part of this step, we select the sites located at the top of the sorted list as hotspots. For that, we pre-specify the hotspots listed size, such as r = 5 percent, 10 percent,... of the n candidate locations.

Step 4. For each site i, simulate accident history according to:

$$Y_i^{sim} \mid \theta_i^{true} \sim Poisson(\theta_i^{true})$$

Step 5. Based on the simulated data Y_i^{sim}, both model parameters and crash risk estimates $\widetilde{\theta}_i$ are computed as follows:

EB approach: Model parameters are first estimated by maximizing the NB marginal distribution and then plugged into Equation (2.7) to obtain the posterior mean crash frequency (according to the Poisson/Gamma model). Parameters are estimated using the MASS software package implemented in R (http://www.r-project.org/).

HB approach: Samples are drawn directly from the posterior distribution of θ_i using MCMC algorithms. Estimate the posterior crash mean for each site using the sample average. This task is developed in the R2WinBUGS software package (Sturtz et al., 2005).

By then sorting the posterior expected accident frequencies of all sites, we obtain the estimated EB and HB ranks:

$$\text{Estimated mean (for each approach): } \{\tilde{\theta}_{[1]}^{est} > \tilde{\theta}_{[2]}^{est} > \ldots > \tilde{\theta}_{[n]}^{est}\}$$
$$\text{Estimated ranks (for each approach): } \{[1], [2], \ldots, [n]\}$$

For each outcome, the r locations placed at the top of the list (r = 5, 10, 15 percent of n) are defined as detected hotspots.

Step 6. Each true hotspot list is then compared with the detected hotspot list derived for each approach (that is, true ranks versus estimated ranks).

For each simulation, we compute two performance measures: the percent deviations and the Spearman correlation coefficient detailed in the following section. To obtain statistically reliable results, each scenario is replicated 100 times.

2.5.2 Performance Evaluation Criteria

We use the following two measures for determining differences in ranks and hotspot identification between two approaches (Miranda-Moreno et al., 2007):

> Percentage deviation: The percent deviation represents the proportion of locations that are different in the two lists of hotspots. For computing the percent deviation, the group of locations under analysis is first ranked according to the true accident estimators, from which the number of sites selected from the top of each list r is designated as the true hotspot list. That is, the n locations are ranked based on the EB or HB approach, from which we select a different hotspot list with r number of elements, referred to as detected hotspots (r is pre-defined as a percent of n). Then we compare each of the detected lists with the true hotspot list. This is done by comparing the percentage of locations that are different between the EB or HB hotspot lists and the true hotspot list. This measure, referred to as the percent deviation, is:

$$\%\ \text{deviation} = 100 \times \left(1 - {}^{s}\!/_{r}\right) \tag{2.8}$$

where s is the number of dangerous locations that are common in two compared lists (for example, EB with true hotspot list) and r is the total number of hotpots selected from the top of a list sorted according to the true accident frequency model (for example, r = 100, 200,... and $r < n$). A high percent deviation is obtained if the EB or HB methods result in a high number of hotspots with respect to the true list.

> Spearman correlation coefficient (ρ): This coefficient is a nonparametric technique that is usually applied to evaluate the degree of linear association between two independent variables. We use this coefficient to measure the degree of association between the true and estimated ranks ordered on the basis of an EB or an HB method. It is computed as follows:

$$\rho = 1 - \frac{6\sum_{i=1}^{n} d_i^2}{n(n^2 - 1)} \tag{2.9}$$

where d_i is the difference between the estimated and true ranks of a specific site i and n is the number of sites to be considered in the simulation (n = 50,100,...). The value of ρ can vary from +1 to −1. A value close to +1 suggests that the estimated vs. true ranks are positively and linearly related. The coefficient ρ can give some important insights into the dimensions of the shifts in the ranking orders of two lists of hazardous sites. Then, combining ρ and percent deviation, we can reach a better conclusion about the differences obtained between these two accident-risk methods.

2.5.3 Data Sets

The analysis presented here is based on data simulated using site-specific attributes from two data sets: highway-railway grade crossings with flashing lights in Canada and four-legged signalized intersections from Toronto, Canada.

Highway-railway intersections with flashing lights: This data set includes 5,067 public highway-railway grade crossings located in Canada that are equipped with flashing lights. At this type of crossing, flashing lights are either post-mounted or cantilevered and are the main active warning devices used to inform highway users of an approaching train. An accident data set containing 5 years of information (1997–2001) is involved in this analysis (Saccomanno et al., 2004). In this data set, the AADT varies between 1 and 57,000 vehicles/day, however, in only five percent of the intersections, the AADT is greater than 10,000. The number of daily trains, which includes freight, passenger, and switching trains, ranges from 1 to 73, but over 50 percent of the crossings have daily traffic of ≤ 5 trains. One of the main characteristics of this accident data set is the low accident frequency per intersection (equal to 0.1 crashes). This results in a large proportion of zero accident counts, which is a typical issue in grade crossing accident data (Hauer and Persaud, 1987; Saccomanno et al., 2004; Oh et al., 2006).

When working with this group of intersections, we consider as the true model parameters the ones calibrated using the whole data set of 5,067 highway/railway intersections with flashing lights. The inverse dispersion parameter ϕ is set to 0.735 and μ_i is given by:

$$\mu_i = (F_{1i} \times F_{2i})^{\beta_0} \exp(\beta_1 + \beta_2 x_{2i} + \beta_3 x_{3i} + \beta_4 x_{4i}) \tag{2.10}$$

where F_{1i}, F_{2i} = Average daily trains and AADT, respectively.

- x_{2i} = arterial/collector
- x_{3i} = train maximum speed
- x_{4i} = whistle prohibition
- $\beta_0 = 0.512$, $\beta_1 = -6.361$, $\beta_2 = 0.018$, $\beta_3 = 0.884$, and $\beta_4 = 0.525$

To observe the effect of sample size on the final output of the analysis, samples of $n = 50, 100, 300$, and so on, crossings are selected randomly from the entire data set.

Four-legged signalized intersections: This data set includes 868 four-legged signalized intersections in Toronto with a total of 54,989 accidents for the period 1990–1995. Contrary to the highway-railway intersections, this data set has a high accident frequency with an observed average of approximately 11.0 accidents per year (see Figure 2.2). Traffic volumes vary widely from approximately 5,300 to 72,300 vehicles/day for main approaches (both directions) and from 52 to approximately 42,600 vehicles/day for minor approaches. Flow ratios, calculated as F_{2i}/F_{1i}, minor over major approach volume, range from 0.2 percent to nearly 100 percent. More details of this data set are provided by Miaou and Lord (2003).

For this study, the true model is defined according to the parameter estimates obtained by Miaou and Lord (2003) for the following model form:

$$\mu_i = \beta_0 (F_{1i} + F_{2i})^{\beta_1} \left(F_{2i} \big/ F_{1i} \right)^{\beta_2} \tag{2.11}$$

where F_{1i} and F_{2i} are the minor and major traffic volumes for the year and $\beta_0 = 0.2001$, $\beta_1 = 1.1950$ and $\beta_2 = 0.3770$. From this model, the estimate reported for the inverse-dispersion parameter, ϕ, is 6.586. For simulation purposes, we use only the traffic volumes for 1995. Again, to assess the impact of sample size, samples of $n = 50$, 100, 300, and so on, sites are selected randomly from the entire data set.

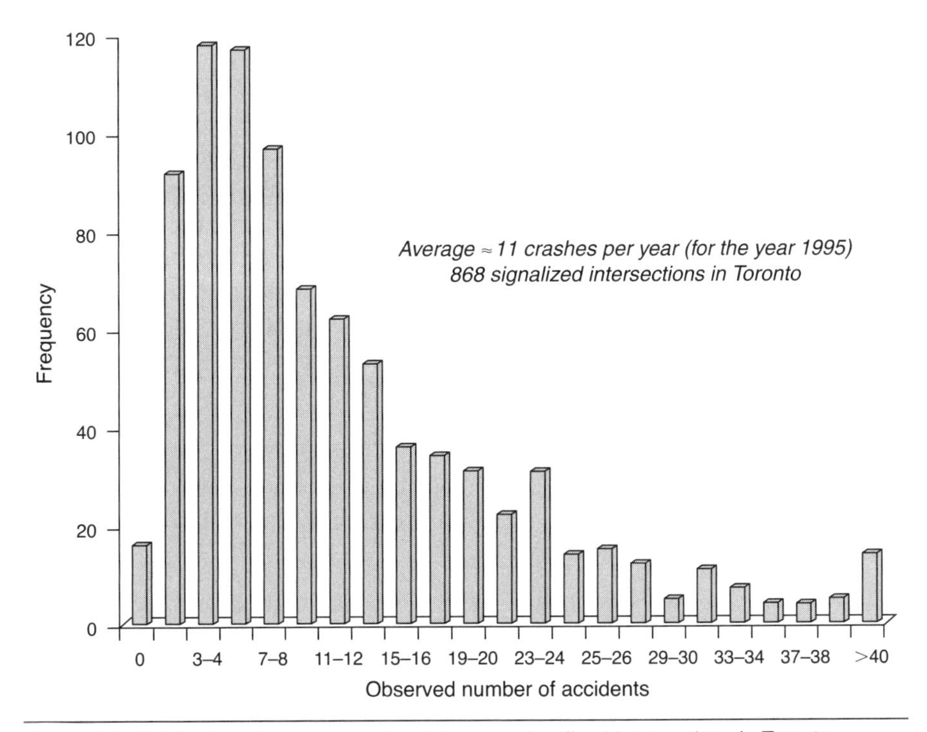

Figure 2.2 Distribution of the observed accidents at signalized intersections in Toronto

2.5.4 Computational Details

The simulated data are first generated using the software R[6]. The posterior quantities of interest are then estimated using WinBUGS 1.4. This software is specialized for the Bayesian analysis of complex statistical models using MCMC methods. Moreover, to make a connection between R and WinBUGS, the R2WinBUGS package is used, providing convenient functions to call WinBUGS from R (Sturtz et al., 2005). On the other hand, when implementing an EB approach, the NB model is calibrated using the MASS package and the glm.nb library in R.

For each data set involved in the analysis, the simulation was replicated 100 times, according to the procedure illustrated in Figure 2.1. At the end of the 100 replications, the standard statistics, such as the average and standard deviation of the percentage deviation and Spearman correlation are computed.

Furthermore, for the MCMC simulations, we specify:

- number of Markov chains = 3
- number of iterations per chain = 8000
- length of burn-in = 4000 (that is, number of iterations to discard at the beginning)
- number of iterations between saving of results = 2000 for each chain (given that the number of thins = 2, which represents the thinning rate). Thus, 6000 iterations were used for parameter estimations.

For the hierarchical Poisson/Gamma model, noninformative priors are specified for the vector of regression parameters (β), that is, $\beta_j \sim N(0, 1000)$. On the contrary, a semi-informative prior is assumed on ϕ by fixing the hyper-parameters a and b according to the ϕ-estimate value obtained by maximizing the NB marginal likelihood. For instance, for the highway/railway crossing data set, the NB model was first calibrated, yielding an inverse dispersion parameter of $\phi = 0.735$. Based on the expectation of the Gamma hyper-prior for ϕ is a/b, and by fixing $a = 1$, it can be assumed that $E(\phi|b) = 1/b = = 0.735$, from which $b = 1.4$.

2.6 RESULTS

The simulation results for various sample sizes are summarized in Tables 2.1-2.4. Based on these results, we observe:

- When using the Toronto data set, the percent deviations and Spearman correlation coefficients are practically the same for the EB and HB estimators involved in this analysis (Table 2.1). The average values from the 100 simulations are, in general, comparable and consistent for the sample sizes (n=50, 100, 300, and so on observations). This suggests that when working with data sets with high accident counts, the differences between the EB and HB approaches can be small. In other words, the hotspot identification process seems to be sensitive to neither the approach (EB or HB) nor sample size when working with these types of data sets. This result supports the finding of Leyland and Davies (2005) who argued that when count data are extensive (large samples), the results from two approaches are consistent.

Table 2.1 EB vs. HB: Toronto intersections

n = 50 intersections						
	% deviation*					Ranking correlation (ρ)
Approach	$r = 5$	$r = 10$	$r = 15$	$r = 20$	$r = 30$	
EB	0.25	0.20	0.18	0.16	0.15	0.93
HB	0.25	0.20	0.19	0.16	0.15	0.93
n = 100 intersections						
	% deviation*					Ranking correlation (ρ)
Approach	$r = 5$	$r = 10$	$r = 15$	$r = 20$	$r = 30$	
EB	0.21	0.19	0.18	0.17	0.15	0.92
HB	0.21	0.19	0.18	0.17	0.15	0.92
n = 300 intersections						
	% deviation*					Ranking correlation (ρ)
Approach	$r = 5$	$r = 10$	$r = 15$	$r = 20$	$r = 30$	
EB	0.20	0.17	0.16	0.14	0.13	0.94
HB	0.20	0.17	0.16	0.14	0.13	0.94

*The percent deviation was computed for different values of r which is given in percent of n, (r is the number of hotpots selected from the top of a list). All the average values are based on the 100 simulations for each scenario.

- For the highway-railway data set, the percent deviation and Spearman correlation appeared to be sensitive to the sample size (Table 2.2). For instance, when using a sample of n=50 crossings, the HB approach produced slightly better results than the EB approach. Despite semi-informative priors being defined in this analysis, the use of more informative priors is expected to improve estimates when working with few data (see Miranda-Moreno et al., 2007).

Table 2.2 EB vs. HB: Highway-railway crossings

n = **50 crossings**						
Approach	**% deviation***					**Ranking correlation (ρ)**
	$r = 5$	$r = 10$	$r = 15$	$r = 20$	$r = 30$	
EB	0.38	0.42	0.41	0.42	0.40	0.51
HB	0.34	0.37	0.38	0.37	0.37	0.57
n = **100 crossings**						
Approach	**% deviation***					**Ranking correlation (ρ)**
	$r = 5$	$r = 10$	$r = 15$	$r = 20$	$r = 30$	
EB	0.33	0.33	0.32	0.30	0.30	0.65
HB	0.32	0.32	0.30	0.29	0.29	0.66
n = **300 crossings**						
Approach	**% deviation***					**Ranking correlation (ρ)**
	$r = 5$	$r = 10$	$r = 15$	$r = 20$	$r = 30$	
EB	0.33	0.33	0.33	0.32	0.32	0.64
HB	0.33	0.33	0.33	0.32	0.32	0.64

*Average values are based on the 100 simulations for each scenario.

- For the Toronto data set, the MCMC parameter estimates are slightly better than the maximum-likelihood (ML) parameter estimates; however, the differences are relatively small (Table 2.3). In terms of model parameters, these two alternative approaches seem to produce similar results when working with data sets with a high average number of accidents. This is also supported by the recent work of Lord and Miranda-Moreno (2008).
- When working with small sample sizes and low accident counts, MCMC posterior estimates of the inverse dispersion parameter ϕ are much closer to the true value than the ML estimates. For instance, for the case of highway-railway intersections with $n = 50$ (Table 2.4), the average MCMC estimate of ϕ over the 100 simulations is equal to 0.74, which is close to the true value ($\phi^{true} = 0.735$). However, the average ML estimate of ϕ is equal to 15.28 with some cases having extremely high values (and, thus, large standard errors). This is in agreement with the work of Lord (2006) who found that the ML estimators performed poorly under low mean and small sample size.

Table 2.3 Comparison in terms of parameter estimates: Toronto intersections

Sample size	Method	Statistics	True versus estimated parameters			
			$\beta_0 = 0.200$	$\beta_1 = 1.195$	$\beta_2 = 0.377$	$\phi = 6.586$
$n = 50$	ML	Mean	0.240	1.190	0.389	7.763
		St. dev	*0.148*	*0.168*	*0.247*	*2.562*
	MCMC*	Mean	0.239	1.194	0.395	7.222
		St. dev	*0.152*	*0.172*	*0.248*	*1.995*
$n = 100$	ML	Mean	0.228	1.196	0.374	7.152
		St. dev	*0.133*	*0.142*	*0.062*	*1.936*
	MCMC	Mean	0.227	1.202	0.375	6.985
		St. dev	*0.141*	*0.149*	*0.063*	*1.693*
$n = 300$	ML	Mean	0.205	1.200	0.372	6.749
		St. dev	*0.057*	*0.078*	*0.088*	*0.853*
	MCMC	Mean	0.207	1.198	0.374	6.690
		St. dev	*0.060*	*0.080*	*0.088*	*0.831*

*The MCMC estimates correspond to the posterior mean of the model parameter. All the values (mean and standard deviation) are based on the 100 simulations for each scenario.

Table 2.4 Comparison in terms of parameter estimates: highway-railway crossings

Sample size	Method	Statistics	True versus estimated parameters					
			$\beta_0 = -6.361$	$\beta_1 = 0.512$	$\beta_2 = 0.018$	$\beta_3 = 0.884$	$\beta_4 = 0.525$	$\phi = 0.735$
$n = 50$	ML	Mean	−5.78	0.72	0.01	0.27	0.45	15.28
		St. dev	*2.21*	*0.82*	*0.02*	*4.30*	*0.23*	*94.05*
	MCMC	Mean	−6.82	0.88	0.02	0.55	0.53	0.74
		St. dev	*2.81*	*1.17*	*0.03*	*3.89*	*0.28*	*0.29*
$n = 100$	ML	Mean	−6.42	0.47	0.02	0.74	0.53	3.75
		St. dev	*1.50*	*0.54*	*0.01*	*0.96*	*0.17*	*20.26*
	MCMC	Mean	−6.78	0.48	0.02	0.83	0.56	0.91
		St. dev	*1.66*	*0.58*	*0.01*	*1.07*	*0.18*	*0.34*
$n = 300$	ML	Mean	−6.41	0.46	0.02	0.84	0.53	0.84
		St. dev	1.06	0.26	0.01	0.38	0.09	0.25
	MCMC	Mean	−6.53	0.47	0.02	0.86	0.55	0.80
		St. dev	1.09	0.26	0.01	0.39	0.09	0.22

The hotspot identification differences between the EB and HB methods seem to be negligible when working with a large number of sites and/or with a high frequency of accidents, for example, $n > 100$ and $\mu > 10.0$. However, when the data set of interest is characterized by a low mean accident frequency and/or small sample sizes, the HB approach can be more reliable. Furthermore, the HB approach has the advantages of being suitable for (1) adopting easily alternative model structures (for example, space-time hierarchical mixed Poisson or latent class models), (2) computing more complex safety measures and functional forms easily (for example, posterior expectation of ranks), and (3) incorporating prior knowledge in the form of priors obtained from either past studies or practitioner knowledge (Berger, 1985; Carlin and Louis, 2000).

2.7 SUMMARY

This chapter summarizes an investigation on the comparative performance of two alternative Bayesian approaches: EB and HB. Due to its simple mathematical form and intuitive basis, the EB approach has been widely accepted by researchers and practitioners as a rational framework for safety analyses such as hotspot identification and before-after studies. In contrast, the HB approach is more sophisticated mathematically and computationally intensive, but it has the advantage of being more powerful in accounting for all uncertainties in model parameters. With the HB approach, one is also able to compute more complex safety measures and functional forms, and to incorporate prior experiences in the analysis. Moreover, when working with complex accident data, more flexible models can be implemented under an HB framework, including the space-time mixed and mixture Poisson hierarchical models.

In this research, we conducted a simulation study to evaluate the differences in performance between these two Bayesian approaches. Semi-informative priors are assumed on the dispersion parameter for the HB approach. The performance is measured in terms of results of the hotspot identification process in which the two approaches are used to estimate the safety status of the sites under consideration. It was found that the HB method performed better than the EB approach when working with small data sets characterized by a low overall mean accident frequency. However, when the data set is sufficiently large (for example, over 300 sites), the differences between the two approaches seem to be minimal.

In summary, when working with few data, the inclusion of prior knowledge in the dispersion parameter can improve the accuracy of the hotspot identification and parameter estimation process significantly. However, when working with large data sets or noninformative priors, the HB modeling approach via MCMC sampling seems to produce similar parameter and crash risk estimates. As discussed, the use of informative priors is not the sole justification to adopt an HB approach.

The inconvenience is that the HB modeling approach is computationally more time-consuming and complex compared to the EB approach. Therefore, if no prior information or complex models are involved in the traffic safety analysis, the EB approach can be justified.

ACKNOWLEDGMENTS

Support for this study was provided by the Transportation Development Centre of Transport Canada. The opinions, findings, and conclusions expressed in this paper are those of the authors and do not necessarily reflect the views of Transport Canada.

ENDNOTES

[1]Accident occurrence is a rare and random phenomenon. Hence, sites with high accident rates in a given period can experience lower accident rates in a subsequent period, even if no treatment is applied. This means that some ostensible hotspots can result from accident random variation rather than actual safety problems.

[2]Accident data is used first to estimate model parameters, and once parameters are determined, it is used again to make a posterior inference (Hauer, 1997).

[3]Traffic accidents are events with a low probability of occurrence. Each vehicle passing through an intersection or a road segment can be seen as a trial with a low probability of being involved in a traffic accident. In the literature, this process is commonly modeled as a Poisson probability distribution (Lord et al., 2005).

[4]Because of the lack of disaggregate data, traffic exposure is usually defined according to the AADT flows. Little work has focused on the relationships between crashes and other traffic flow characteristics such as vehicle density, level of service, vehicle occupancy, and speed distribution, (Lord et al., 2005b). Few works have considered hourly exposure functions accounting for traffic composition and other temporal variations (Qin et al., 2006).

[5]Instead of assuming a Gamma distribution, alternative error distributions have been suggested in the literature, such as the Log-Normal distribution. The marginal distribution of the Poisson/Lognormal model does not have a closed form. However, to obtain the maximum likelihood estimates, we can use some methods such as the Gauss-Hermite quadrature or the EM algorithm (Winkelmann, 2003).

[6]A language that is specialized for statistical computing is available at http://www.r-project.org

REFERENCES

Abbess, C., D. F. Jarrett, and C. C. Wright. 1981. Accidents at black spots: Estimating the effectiveness of remedial treatment, with special reference to the regression-to-mean effect. *Traffic Engineering and Control*, 22(10): 535–542.

Berger, J. O. 1985. *Statistical Decision Theory and Bayesian Analysis. Springer Series in Statistics*, 2d ed. New York: Springer-Verlag.

Brijs, T., D. Karlis, F. Van den Bossche, and G. Wets. 2007. A Bayesian model for ranking hazardous road sites. *Journal of the Royal Statistical Society: Series A*, 170: 1–17.

Cameron, A. C. and P. K. Trivedi. 1998. *Regression Analysis of Count Data.* Cambridge, UK: Cambridge University Press.

Carlin, B. and V. Louis. 2000. *Bayes and Empirical Bayes Methods for Data Analysis.* Chapman and Hall.

Cheng, W. and S. P. Washington. 2005. Experimental evaluation of hotspot identification methods. *Accident Analysis and Prevention*: 37: 870–881.

Gelman, A., J. B. Carlin, H. S. Stern, and D. B. Rubin. 2003. *Bayesian Data Analysis*, 2d ed. New York: Chapman and Hall/CRC.

Hauer, E. and B. N. Persaud. 1987. How to estimate the safety of rail-highway grade crossings and the safety effects of warning devices. *Transportation Research Record*, 1114: 131–140.

Hauer, E. 1997. *Observational before-after studies in road safety: Estimating the effect of highway and traffic engineering measures on road safety.* Elsevier Science.

Heydecker, B. G. and J. Wu. 2001. Identification of sites for accident remedial work by Bayesian statistical methods: An example of uncertain inference. *Advances in Engineering Software*, 32: 859–869.

Higle, J. L. and J. M. Witkowski. 1988. Bayesian identification of hazardous sites. *Transportation Research Record*, 1185: 24–35.

Leyland, A. H. and C. A. Davies. 2005. Empirical Bayes methods for disease mapping. *Statistical Methods in Medical Research*, 14: 17–34.

Lord, D., S. P. Washington, and J. N. Ivan. 2005. Poisson, Poisson-Gamma and Zero-inflated regression models of motor vehicle crashes: Balancing statistical fit and theory. *Accident Analysis and Prevention*: 37(1): 35–46.

Lord, D., A. Manar, and A. Vizioli. 2005b. Modeling crash-flow-density and crash-flow-v/c ratio for rural and urban freeway segments. *Accident Analysis and Prevention*, 37(1): 185–199.

Lord, D. 2006. Modeling motor vehicle crashes using Poisson-gamma models: Examining the effects of low sample mean values and small sample size on the estimation of the fixed dispersion parameter. *Accident Analysis and Prevention*, 38(4): 751–766.

Lord, D. and L. F. Miranda-Moreno. 2008. Effects of low sample mean values and small sample size on the estimation of the fixed dispersion parameter of Poisson-gamma models for modeling motor vehicle crashes: A Bayesian perspective. *Journal of Safety Science*, 46: 751–770.

Lord, D., S. Guikema, and S. Geedipally. 2008b. Application of the Conway-Maxwell-Poisson Generalized Linear Model for Analyzing Motor Vehicle Crashes. *Accident Analysis & Prevention*, 40(3): 1123–1134.

Miaou S. P. and D. Lord. 2003. Modeling traffic crash-flow relationships for intersections: Dispersion parameter, functional form, and Bayes versus empirical Bayes. *Transportation Research Record*, 1840: 31–40.

Miaou, S. P. and J. J. Song. 2005. Bayesian ranking of sites for engineering safety improvement: Decision parameter, treatability concept, statistical criterion and spatial dependence. *Accident Analysis and Prevention*, 37: 699–720.

Miranda-Moreno, L., L. Fu, F. Saccomano, and A. Labbe. 2005. Alternative risk models for ranking locations for safety improvement. *Transportation Research Record*, 1908: 1–8.

Miranda-Moreno, L. F., A. Labbe, and L. Fu. 2006. Multiple Bayesian testing procedures for hotspot identification. *Accident Analysis and Prevention*, 39(6): 1192–1201.

Miranda-Moreno, L. F. 2006. "Models and methods for identifying hazardous locations for safety improvements." PhD thesis, University of Waterloo, Canada.

Miranda-Moreno, L. F., D. Lord, and L. Fu. 2007. "Evaluation of alternative hyper-priors for Bayesian road safety analysis." Paper presented at the Annual Meeting of the Transportation Research Board, Washington, DC.

Oh, J. S., S. Washington, and D. Nam. 2006. Accident prediction model for railway-highway interfaces. *Accident Analysis and Prevention*, 38(1): 870-881, 346-356.

Qin, X., J. N. Ivan, and N. Ravishanker. 2004. Selecting exposure measures in crash rate prediction for two-lane highway segments. *Accident Analysis and Prevention*, 36(2): 183–191.

Qin, X., J. N. Ivan, N. Ravishanker, R. Liu, and D. Tepas. 2006. Bayesian estimation of hourly exposure functions by crash type and time of day. *Accident Analysis and Prevention*, 38(6): 1071–1080.

Persaud, B., C. Lyon, and T. Nguyen. 1999. Empirical Bayes procedure for ranking sites for safety investigation by potential for safety improvement. *Transportation Research Record*, 1665: 7–12.

Saccomanno, F. F., L. Fu, and L. F. Miranda-Moreno. 2004. Risk-based model for identifying highway-rail grade crossing black spots. *Transportation Research Record*, 1862: 127–135.

Schluter, P. J., J. J. Deely, and A. J. Nicholson. 1997. Ranking and selecting motor vehicle accident sites by using a hierarchical Bayesian model. *The Statistician*, 46(3): 293–316.

Song, J. J., M. Ghosh, S. Miaou, and B. Mallick. 2006. Bayesian multivariate spatial models for roadway traffic crash mapping. *Journal of Multivariate Analysis*, 97: 246–273.

Sturtz, S., U., Ligges, and B. Gelman. 2005. R2WinBUGS: A package for running WinBUGS. *Journal of Statistical Software*, 12(3): 1–16.

Rao, J. N. 2003. *Small area estimation*. New York: John Wiley and Sons.

Tunaru, R. 2002. Hierarchical Bayesian models for multiple count data *Austrian Journal of Statistics*, 31(3): 221–229.

Winkelmann, R. 2003. *Econometric Analysis of Count Data*. Heidelberg, Germany: Springer-Verlag.

3

UTILIZING DATA WAREHOUSE TO DEVELOP FREEWAY TRAVEL TIME RELIABILITY STOCHASTIC MODELS

Haitham M. Al-Deek
Department of Civil & Environmental Engineering
University of Central Florida

Emam B. Emam
URS Corporation

3.1 INTRODUCTION

Although travel time reliability has not been a common performance measure, it is important for providing travelers with accurate route guidance information. A traffic data warehouse with extensive amounts of loop detector traffic counts and vehicle speeds was used to develop a new methodology for estimating travel time reliability of Interstate 4 (I-4) corridor in Orlando, Florida. Four travel time stochastic models: Weibull, Exponential, Lognormal, and Normal were investigated. The developed best-fit stochastic model (Lognormal) can be used to estimate and predict travel time reliability of freeway corridors and report this information in real time to the public through traffic management centers. Analysis of the four-month data demonstrated that using long segment lengths (for example, 25 miles) in calculating travel time reliability can be misleading. It is recommended that shorter segment lengths on the order of 5 miles be used in calculating travel time reliability for freeway corridors. The new method was compared with existing methods (Florida and Buffer Time (BT) methods). The vulnerability of existing methods in estimating reliability of congested segments with high travel time variability was demonstrated using warehouse data. Unlike the existing methods, which were insensitive to the travelers' perspective of travel time, the new method showed high sensitivity to the geographical location that reflects the level of congestion and bottlenecks. The major advantage of the new method over existing methods to practitioners and researchers is its ability to estimate travel time reliability as a function of departure time. As such, it is more appropriate for measuring the performance of freeway operations.

During the twentieth century, transportation programs focused on the development of basic infrastructure of transportation networks. In the twenty first century, however, the focus has shifted to management and operations of these networks (NCHRP, 2003). Reliability measures will now be playing a more important role in judging the performance of the transportation system and in evaluating the impact of new Intelligent Transportation Systems (ITS) deployment. Transportation systems may even rely on current and projected trends in reliability measures to assess transportation deficiencies and potential improvements to set funding or programming priorities. This can be done by comparing competing alternatives to achieve up-to-date, reliable travel-time information to the commuting public based on current or historic databases.

Network reliability theory has been applied extensively in many real-world systems such as computer and communication systems, power transmission, and distribution systems. In this context, Kececioglu (1991) defined reliability as "the probability that an entity will perform its intended function(s) without failure for a specified length of time under the stated operating conditions at a given level of confidence."

In the transportation field, network reliability has been the subject of considerable international research interest (see Bell and Cassir, 2000 and Bell and Iida, 2003). The analysis of network reliability involves measuring the ability of the network to meet some expected functional criteria or tolerance set by its users, which in turn varies according to the severity and frequency of the underlying recurrent and nonrecurrent causes. Recurrent is generally characterized by everyday rush hour stop-and-go conditions that occur when demand exceeds capacity. Nonrecurrent causes flow from incidents, maintenance, construction, or special events that cause higher-than-normal peak demand.

Travel time reliability is challenging to achieve because there is no single agreed-upon travel time reliability measure. Reliability has been defined as the probability that a trip between a given origin-destination pair can be successfully made within a specified time interval (Asakura and Kashiwadani, 1991; Chen et al., 1999; Yang et al., 2000; Levinson and Zhang, 2001). These researchers calculated the probability that the travel time will be within a predefined interval (lower and upper bounds). Chen and Recker (2000) assumed travel time reliability as a function of the ratio of the travel times under the degraded and nondegraded states.

This chapter defines travel time reliability in a totally different manner from existing methods. Unlike those that define reliability in terms of travel time variability, the definition of reliability in the new method puts more emphasis on the users' perspective. The new method has the potential for estimating travel time reliability as a function of departure time. This is accomplished by treating travel time as a continuous variable that captures the variability experienced by individual travelers over an extended period.

This chapter presents a literature review on estimating travel time reliability, a review of the four stochastic models techniques used, the experimental design of the study, including data collection and preparation, and an evaluation of the developed models compared with the existing methods. It concludes with contributions to transportation engineering, recommendations and suggested future research.

3.2 LITERATURE REVIEW

This section is an overview of a number of techniques that have been used in the United States to measure travel time reliability in the transportation field.

3.2.1 Percent Variation

Statistically, *percent variation* is known as the *coefficient of variation* (CV) and is calculated as the standard deviation divided by the mean. A traveler can multiply average travel time by the percent variation and then add that product to average trip time to get the time needed to be on time about 85 percent of the time (Turner et al., 1996). This measure describes the variability of travel time, but it falls short of explaining how conditions on the corridor meet travelers' expectations (Shaw and Jackson, 2003).

3.2.2 California Reliability Method

California Transportation Plan (1998) defines *reliability* as the variability between the expected travel time based on scheduled or average travel time and the actual travel time caused by the effects of nonrecurrent congestion (Lomax et al., 2004). The coefficient of variation of average trip time distribution is used as the reliability index. Segments with insignificant travel time variations from day to day have narrower curves of average trip times and are considered reliable. Because this measure reflects the variation in travel times more than the acceptability of the travel times to the user, it is incomplete.

3.2.3 Florida Reliability Method

The Florida Reliability Method was derived from the Florida Department of Transportation (FDOT) definition of reliability of a highway system as the percent of travel (trips) on a corridor that takes no longer than the expected travel time (the median travel time across the corridor during the period of interest) plus a certain acceptable additional time. This additional acceptable time is a percentage of the expected travel time and is used to establish the additional time beyond the expected travel time a traveler finds acceptable. Percentages of 5, 10, 15, and 20 percent above the expected travel time have been considered, but without a final decision (Shaw and Jackson, 2003). Mathematically, reliability $R(t)$ is defined (FDOT 2000) as the probability of travel that is no longer than an acceptable travel time, TT, as follows:

$$R(t) = P(x < X + \Delta) = P(x < TT)$$

where:

$X \rightarrow$ The median travel time across the corridor during the period of interest

$\Delta \rightarrow$ A percentage of the median travel time during the period of interest

3.2.4 Buffer Time

The buffer time concept may relate particularly well to the way travelers make decisions. It uses minutes of extra travel time needed to allow on-time arrival (Chen et al., 2003). The buffer time is the difference between the average and the upper limit of the 95 percent confidence interval as calculated from the annual average. The main problem is that the public does not readily understand buffer time. Mathematically, it is calculated as follows:

$$\text{Buffer time} = \left[\begin{matrix}\text{95\% Confidence Travel Time for} \\ \text{an Average Trip (in minutes)}\end{matrix} - \begin{matrix}\text{Average Travel Time} \\ \text{(in minutes)}\end{matrix}\right]$$

(Chen et al., 2003)

To track reliability measures, buffer time and Florida reliability methods appear to be the most reasonable of the preferred measures, and they also seem to resonate with most audiences. However, for these measures it is not certain what level of reliability (85, 90, or 95 percent) should be used. Most of the previous studies have had a theoretical rather than an empirical basis for model development. None of the previous studies compare the most common existing methods using warehouse data. In addition, none of the existing methods consider the impacts of segment length on travel time reliability calculations and accuracy. As such, the main goal is to develop a methodology that addresses the major shortcomings.

Specific objectives include the development of an efficient methodology that can predict the travel time reliability of freeway corridors based on real-time and historic loop-detector data. Four statistical stochastic models are investigated to discover the best-fit probability density function using available travel time data. Travel time reliability is then calculated using the best-fit stochastic model and the engineering reliability models that have not been used before in the transportation field. The best-fit travel time reliability model is implemented on a section of I-4 in Orlando, Florida. This section carries large volumes of long distance, interregional, and intrastate trips.

3.3 RESEARCH FRAMEWORK

This study investigates models to forecast travel time reliability, including testing the transferability of reliability techniques to the transportation field, and focuses on four stochastic models. The research conceptual plan, experimental design, and traffic data used are explained and then the results and applications are discussed.

3.3.1 Research Conceptual Plan

The literature reveals the need to develop a new methodology for modeling travel time reliability which utilizes the massive traffic data available in freeway data warehouses. Figure 3.1 shows the conceptual plan for modeling freeway travel time reliability. An experimental design was developed to find the best-fit travel time stochastic model (Weibull, Exponential, Lognormal, and Normal). Anderson-Darling (AD) and the 95th percentile of the absolute error were two evaluation criteria selected to check the validity of assumptions regarding the specific stochastic model parameters.

An adjustment of the best-fit stochastic model location parameter is needed since freeway segments cannot have zero travel time. Next, if the calculated/expected travel time (see Equation (3.1)) for a specific segment on a specific day is less than the acceptable travel time, then the segment is considered 100 percent reliable. Here, the acceptable travel time is defined as travel time at speed limit. If the segment's travel time is less than or equal to the free-flow travel time or travel time at the posted speed limit, then this segment is considered 100 percent reliable. Otherwise, the developed reliability stochastic models is used to calculate the segment/corridor travel time reliability as a function of the departure time.

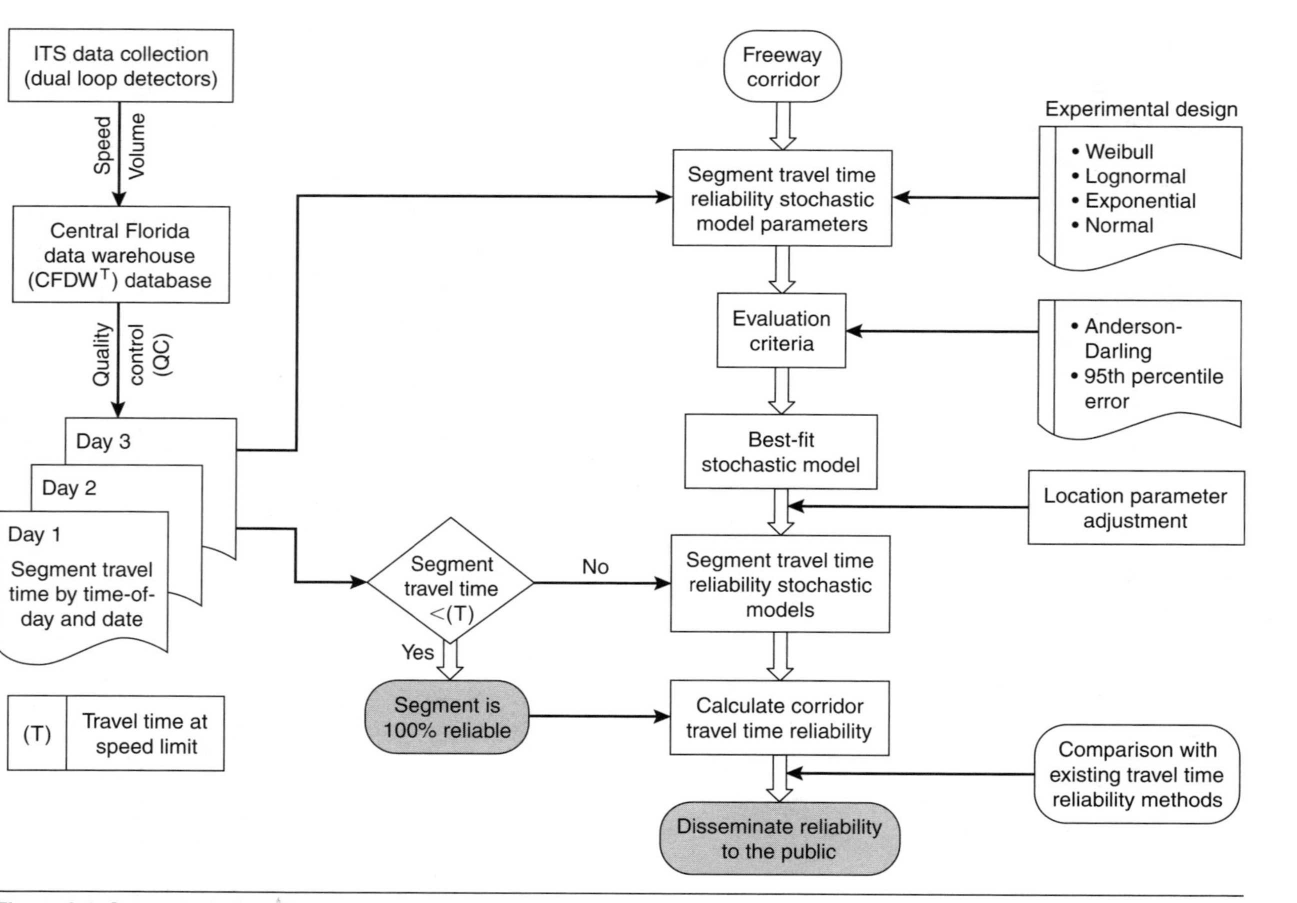

Figure 3.1 Conceptual plan

The new method results are compared with existing FDOT and buffer time (BT) methods. Then the travel time reliability estimates are disseminated to the public in real time via advanced traveler information systems to help in trip planning and possible need for enroute diversion.

3.3.2 I-4 Traffic Data Warehouse

The Orlando Regional Transportation Management Center (RTMC) monitors 50 miles of I-4 corridor. This research studied a 25-mile long section of this corridor (eastbound direction only) (see Figure 3.2). There are dual loop detectors placed approximately 0.5 miles apart. The loop detectors collect traffic volume, lane detector occupancy, and speed data. The data from these detectors are sent automatically to the RTMC every 30 seconds in a binary format and converted to an ASCII-text format. The University of Central Florida (UCF) used the loop detectors data to update the interactive geographic information systems (GIS) speed map that UCF developed on its Internet site (http://www.iflorida.org).

Travel times are derived directly from the speed by considering two consecutive loop detector stations that provide spot-speed measurements. It is assumed that the spot-speed estimate is valid half the distance to the adjacent loop detector. The origins and destinations are the major interchanges (from on-ramp to off-ramp end points) where traffic conditions are likely to change. Therefore, travel time between any on- and off-ramp combination (origin-destination) can be calculated as follows (William and Laurence, 2002):

$$TT = A + \sum_{i=2}^{n-2} \left(\frac{l_{i,i+1}}{2S_i} + \frac{l_{i,i+1}}{2S_{i+1}} \right) + B \tag{3.1}$$

where:

$$A = \begin{cases} \dfrac{X_1}{S_2} & \text{If} \quad X_1 \le \dfrac{l_{1,2}}{2} \\[2ex] \dfrac{l_{1,2}}{2S_2} + \dfrac{\left(X_1 - \dfrac{l_{1,2}}{2} \right)}{S_1} & \text{Otherwise} \end{cases}$$

$$B = \begin{cases} \dfrac{X_{n-1}}{S_{n-1}} & \text{If} \quad X_2 \le \dfrac{l_{n-1,n}}{2} \\[2ex] \dfrac{l_{n-1,n}}{2S_{n-1}} + \dfrac{\left(X_2 - \dfrac{l_{n-1,n}}{2} \right)}{S_n} & \text{Otherwise} \end{cases}$$

i = Detector station i
$l_{i,i+1}$ = Distance between detector station i and $i + 1$
S_i = Spot speed at detector station i
n = Number of detector stations
X_1 = Distance upstream of the first detector in the set
X_2 = Distance downstream of the last detector in the set
TT = Travel time

With any study that includes a large amount of data, the first step in analysis is to determine which data are useful to the study. For example, the temporal means (standard deviations) of travel times from station 5 to station 52 (or from E. US 192 to E. SR 414) on Tuesday, October 21, 2003, are 39.91 minutes (15.22 minutes) and 33.94 minutes (3.86 minutes) for raw and imputed data respectively. Similar results were obtained from the analysis of the rest of the data. Accordingly, it can be concluded that the imputed data reveals better performance, especially during the periods in which the raw data are missing or have abnormal values because such outliers could generate biased mean travel time. Weekdays with observed loop detector raw data (30-second intervals) less than 75 percent of the total daily corridor data were excluded from the analysis. This condition was implemented to exclude days with a significant amount of missing data and to avoid bias in the calculation of mean travel time.

Only imputed data were used in the modeling process. Travel time between any on and off ramp on I-4 was calculated (see Equation (3.1)) using the real-time filtered loop speeds stored in the I-4 data warehouse during 2003 (Al-Deek et al., 2004; Al-Deek and Chilakamarri, 2004).

3.3.3 Data Preparation

A software program was applied to obtain the data of four consecutive months (fall season) from September 1 to December 31, 2003. Data were organized into individual files, each containing 24 hours of information. For the four months considered during the evening peak hours (no holidays and no incidents), there were 50 weekdays valid for analysis in this study. Each weekday had at least 175,000-loop detector records or 75 percent of the available daily raw data. Preliminary analysis indicated that weekends had a different peak period, and that the number of weekdays with incidents during the three peak hours was insignificant (only two). Therefore, this study focused on weekdays with no incidents with a sample of 50 weekdays.

In this study, an origin started at loop detector ID_5 (on ramp, E. US 192) and the set of destinations (off ramps) were labeled to the closest loop detector ID (that is, 14, 22,..., 52) as shown in Figure 3.2. Using Equation (3.1), the travel time standard deviations and means were calculated every 5 minutes during the evening peak period (3:30 PM to 6:30 PM) for the eastbound direction.

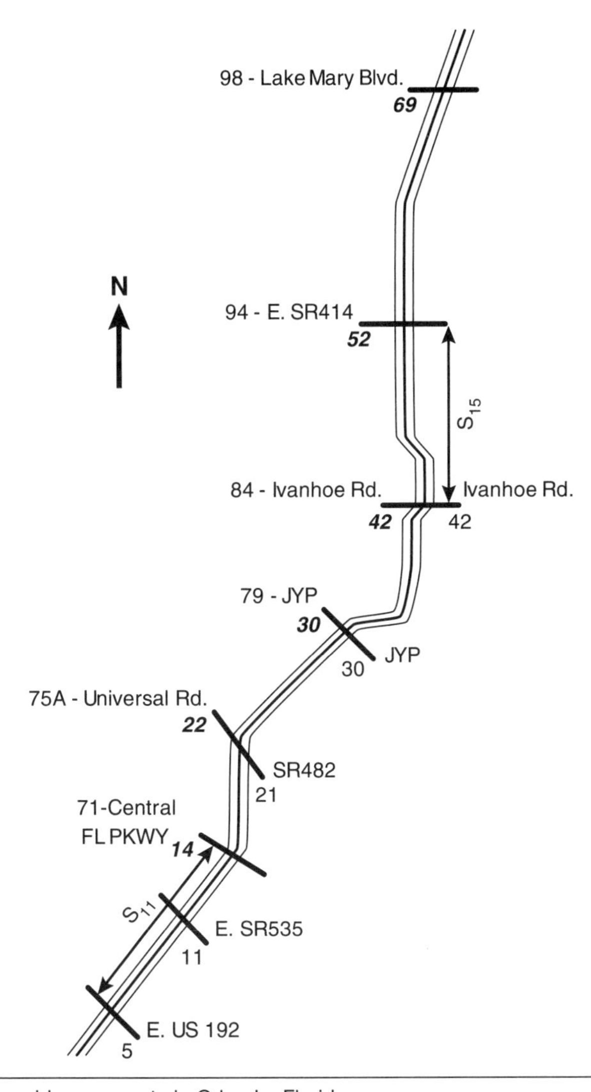

Figure 3.2 I-4 corridor segments in Orlando, Florida

3.3.4 Potential Independent Variables

The main goal of a forecasting model is to emulate a relationship between the dependent and independent variables. The dependent variable is the travel time. Numerous independent variables were possible from the large amount of data recorded in the I-4 data warehouse.

The independent variables describe the stochastic model (probabilistic distribution type), day of the week, month of the year, and location on I-4. The first independent variable is the stochastic model with the four possible variables: Normal, Exponential, Weibull, and Lognormal. We need to find which has the best fit with real data. The next independent variable is the traveled distance on I-4, which extends from E. US 192 (Station 5) to E. SR 414 (Station 52). This section was divided into a range of segments of approximately 5, 10, 20, and 25 miles in length, as shown in Table 3.1. In this table, S_{ij} represents the I-4 segments in which (i = 1, 2, 3, and 4) refers to the approximate length of the segment (that is, 5, 10, 20, and 25 miles) respectively, and (j = 1, 2, 3, 4, and 5) refers to the segment location on I-4 with respect to the loop detector ID_5 (on ramp, E. US 192). For example, S_{11} refers to the first segment of the 5-mile segments group that starts at loop detector ID_5 (on ramp, E. US 192) and ends at loop detector ID_14 (off ramp, 71-Central Florida Parkway) as shown in Table 3.1 and Figure 3.2. While S_{15} refers to the last segment of the 5-mile segments group that starts at loop detector ID_42 (on ramp, Ivanhoe Rd.) and ends at loop detector ID_52 (off ramp, 94-E. SR 414) and so on. The next variable is the *day of the week*. This variable had five possible weekdays (Monday through Friday), which can be explained as follows: The first level *M* included data from Mondays of the four months analyzed in 2003; the second level *T* included data from Tuesdays of the four months analyzed in 2003, and so on. The last independent variable was the *month* under consideration in the analysis sample. This variable had four possible levels (Sep., Oct., Nov., Dec.). These four levels explored the effect of mixing data in the same sample from different weekdays within the same month.

Table 3.1 Analysis segments

Group (length)	Segment	Origin (Loop ID)	Destination (Loop ID)	Distance (miles)
G1 (5 mi)	S11	5	14	5.71
	S12	11	22	6.38
	S13	21	30	4.95
	S14	30	42	5.19
	S15	42	52	4.83
G2 (10 mi)	S21	5	22	10.14
	S22	21	42	10.14
	S23	11	30	10.84
	S24	30	52	10.02
G3 (20 mi)	S31	5	42	19.79
	S32	11	52	20.86
G4 (25 mi)	S41	5	52	24.62

3.3.5 ANOVA Test of Significance

The independent variables were identified from the available I-4 data warehouse. It is possible that some of the independent variables are not significant or highly-related. Thus, it is necessary to perform statistical significance tests on the independent variables using ANOVA for the proposed dependent variable *travel time*. The null and alternative hypotheses being tested are (Mendenhall and Sincich, 1995; Rencher, 2002):

$$H_o: \mu_1 = \mu_2 = \ldots = \mu_i$$
$$H_a: \text{at least two of the } \mu_I\text{s are different}$$

An ANOVA table was applied to each independent variable. For the *time of day* independent variable, the ANOVA test compared the five samples (Monday through Friday) to determine if the underlying mean travel time of each sample was the same (null hypothesis) or significantly different (alternative hypothesis). The ANOVA table was applied to each independent variable and the corresponding F-statistic and p-values are given in Table 3.2. The ANOVA analysis shows that all of the independent variables are significant ($P < 0.0001$). As a result, the possible combination of scenarios of the above variable levels is [$4 \times (5 + 4) = 36$] different scenarios (4 *stochastic models*, 5 *days of the week*, and 4 *months* of the year) for each freeway segment.

Table 3.2 Independent variable significant test results

	Month of the year (Sep, Oct, Nov, & Dec)		Day of the week (M, T, W, TH & F)	
Segment	***F*-Value**	**Pr > *F***	***F*-Value**	**Pr > *F***
S11	198.48	<.0001	37.42	<.0001
S12	37.06	<.0001	46.83	<.0001
S13	24.08	<.0001	114.74	<.0001
S14	40.29	<.0001	172.43	<.0001
S15	13.24	<.0001	7.54	<.0001
S21	119.41	<.0001	48.59	<.0001
S22	39.52	<.0001	206.97	<.0001
S23	25.31	<.0001	123.47	<.0001
S24	46.20	<.0001	148.99	<.0001
S31	43.22	<.0001	214.76	<.0001
S32	44.46	<.0001	184.84	<.0001
S41	294.23	<.0001	186.33	<.0001

3.4 SEGMENT TRAVEL TIME RELIABILITY STOCHASTIC MODELS

The next step in the study framework is to review potential stochastic models for estimating travel time reliability. Four stochastic models are considered.

The key to estimating travel time reliability as a performance measure is in defining the segment travel time probability density function (stochastic model). Once the travel time distribution of a segment is known, the reliability function of that segment can be determined. The statistical analysis software (SAS) software program is used to calculate the distribution parameters and the evaluation criteria. The following subsections briefly demonstrate distribution parameters and the corresponding reliability functions.

3.4.1 Weibull Stochastic Model

The Weibull model is a general-purpose reliability stochastic model. Because of its flexible shape and ability to model a wide range of failure rates, Weibull has been used successfully in many applications as a purely empirical model (Kececioglu, 1991; Tobias and Trindade, 1995). In its most general case, the three-parameter Weibull probability density function (pdf) is defined by:

$$f(T_i)=\frac{\beta}{\eta}\left(\frac{T_i-\gamma}{\eta}\right)^{\beta-1} e^{-\left(\frac{T_i-\gamma}{\eta}\right)^{\beta}} \tag{3.2}$$

where $f(T_i) \geq 0$, $T_i \geq \gamma$, $\beta > 0$, $\eta > 0$, $-\infty < \gamma < \infty$, T_i is the segment travel time, β is defined as the shape parameter, η is the scale parameter, and γ represents the model location parameter. The Weibull reliability function $R(T_i)$ as a function of the unreliability function $F(T_i)$ is:

$$R(T_i)=e^{-\left(\frac{T_i-\gamma}{\eta}\right)^{\beta}}=1-F(T_i) \tag{3.3}$$

3.4.2 Exponential Stochastic Model

The exponential model is a special case of the Weibull model with shape parameter $\beta = 1$. If the travel times of a segment are well represented by the exponential model with a scale parameter ($\eta = 1/\lambda$) and a mean failure rate λ, then using Equation (3.3), the travel time reliability function can be written as:

$$R(T_i)=e^{-\lambda(T_i-\gamma)}=1-F(T_i) \tag{3.4}$$

3.4.3 Lognormal Stochastic Model

The lognormal stochastic model has a flexible probability density and failure rate functions. It has a variety of shapes that can resemble the shapes of the Weibull model (Kececioglu, 1991; Tobias and Trindade, 1995). This flexibility makes the lognormal an empirically useful model for right-skewed data. The pdf of the three-parameter lognormal distribution λ, T_{50}, and $\sigma_{T'}$ is given as:

$$f(T_i) = \frac{1}{(T_i - \gamma)\sigma_{T'}\sqrt{2\pi}} e^{-\frac{1}{2}\left(\frac{Ln(T_i-\gamma)-\overline{T}'}{\sigma_{T'}}\right)^2} \tag{3.5}$$

where:

$$f(T_i) \geq 0, T_i \geq \gamma, -\infty < \overline{T}' < \infty, \sigma_{T'} > 0, -\infty < \gamma < \infty$$
$$T' = Ln(T_i) \rightarrow \text{ where } T_i \text{ is the travel time}$$

The model parameters are: location parameter (γ), scale parameter $(\overline{T}' = Ln(T_{50}))$, and shape parameter ($\sigma_{T'} = \sigma$). Then, the reliability function can be written as:

$$R(T_i) = 1 - F(T_i) = 1 - \Phi\left\{\frac{Ln(T_i - \gamma) - Ln(T_{50})}{\sigma}\right\} \tag{3.6}$$

3.4.4 Normal Stochastic Model

The normal distribution is the logarithm of the lognormal model with location parameter ($\gamma = 0$), mean or scale parameter ($\mu = LN(T_{50})$), and standard deviation or shape parameter ($\sigma = \sigma_{T'}$). The reliability function can be written as:

$$R(T_i) = 1 - F(T_i) = 1 - \Phi\left\{\frac{T_i - \hat{\mu}}{\hat{\sigma}}\right\} \tag{3.7}$$

3.4.5 Goodness-of-Fit Tests

To check the validity of our assumptions regarding the specific travel time stochastic model, the goodness-of-fit (GoF) tests that are based on either the cumulative distribution function (CDF) such as Anderson-Darling (AD) and the Kolmogorov-Smirnov (KS) tests or the pdf such as the Chi square test can be used. We have selected AD because it is among the best distance tests for small and large samples, and various statistical packages are widely available for it (Shimokawa and Liao 1999). The AD statistic is a weighted squared distance from the plotted points to the fitted line with larger weights in the tails of the distribution with the null hypothesis:

H_o: The data followed the specified distribution
H_a: The data did not follow the specified distribution

The 95th percentile of the absolute error, which was calculated by comparing the predicted and the actual model distribution, is also used as another evaluation criterion to assess the assumed model parameters.

3.4.6 System Travel Time Reliability

In a series configuration, as in the freeway corridor, a failure of any individual segment causes an overall system failure. Readers are referred to our paper (Al-Deek and Emam, 2006) for more details. Mathematically, the reliability of the corridor is given by:

$$R_S = P(X_1)P(X_2)...P(X_n) = \prod_i^n P(X_i) = \prod_i^n R(X_i) = \prod_i^n R_i \quad (3.8)$$

where:

R_S = reliability of the system and
$P(X_i)$ = the probability that segment i is operational

For example, if all the components have the same reliability $R(x) = 0.95$, then the system reliability of a series configuration of $n = 10$ components will be:

$$R_S = \prod_i^n R_i = 0.95^{10} \approx 0.60$$

hence, the system reliability of a series configuration is much lower than 0.95. Now, we move to the model application of this developed framework to the I-4 corridor and data warehouse previously described in the framework.

3.5 MODEL APPLICATION

Using this I-4 data warehouse we are able to estimate the travel time stochastic model parameters. Once the model with the best fit of the segment was selected an adjustment was needed; one that must reflect that 100 percent travel time reliability does not start right at zero travel time. It is unrealistic that any highway segment can be traversed in zero seconds. The public also has a perception of an acceptable travel time threshold at which 100 percent travel time reliability is achieved. The logical assumption is to consider the road segment 100 percent reliable if its travel time is less than or equal to an upper threshold. We chose this upper threshold so that each segment would be equal to the segment length divided by the speed limit. This assumes the public will be completely satisfied if traffic conditions allow them to travel the roadway facility at the speed limit, an assumption that can be verified by surveys.

3.5.1 Segment Travel Time Distribution

The SAS was used to fit the data to the theoretical travel time stochastic models for the 12 analyzed segments. These segments were categorized based on their mean travel speed as uncongested segments (for example, S_{11}, S_{12}, and S_{21}) and congested segments (for example, S_{13}, S_{14}, S_{15}, S_{22}, S_{23}, S_{24}, S_{31}, S_{32}, and S_{41}). The two congestion categories were classified based on the variation in loop detectors speed and how close their observed speeds were to the speed limits during the peak hours.

Once the stochastic models were developed, the AD goodness-of-fit test was calculated to test the assumption that the travel time follows the four tested models. Figure 3.3 shows the percentages of the fitted models that have AD values less than 1.5. It is obvious that as long as data was gathered from different days of different months, the estimation of parameters got worse as evidenced by the large AD statistic, which has a negative impact on estimation accuracy. The null hypothesis is rejected at 99 percent confidence level for all AD values exceeding the appropriate critical value (1.5) as in D'Agostino and Stephens (1986). Figure 3.3 shows that on average, 50.0 percent or more of the lognormal models had AD values less than the critical value (1.5) for the weekday data. The aforementioned value dropped to 16.7 percent for samples with mixed weekdays, such as September, and reached 0.0 percent for the *all days* sample (50 weekdays). The results demonstrated that using data for the same weekdays yielded better results than mixing data for different weekdays within the same month or from different months in the same sample. The exponential model failed to meet this criterion for all scenarios.

The second criterion was the estimated error percentages which were calculated by comparing the predicted and the actual distributions. Among the analyzed scenarios, the 95th percentile of the absolute errors of the lognormal models were less than 5.78 percent compared to 9.03, 14.10, and 23.52 percent of the Weibull, normal, and exponential models respectively (see Figure 3.3). Consequently, the lognormal proved to be the best stochastic model that fits travel time, especially for the day of the week samples (Monday through Friday). This was followed by the Weibull distribution. The exponential distribution had the highest 95th percentile (23.53 percent) and the highest AD values (none were below the 1.5 threshold), so it was ranked lowest.

Based on these results, the rest of the paper analysis only focuses on the first five categories (or five levels) of data that represent weekdays (Mondays, Tuesdays, ..., and Fridays). This leads to 60 lognormal model scenarios (5 weekdays $\times$ 12 segments).

3.5.2 Spatial Correlation in Travel Times

Based on observations of the travel time data available in the UCF data warehouse, it is inappropriate to assume that segment travel times are generally independent. For

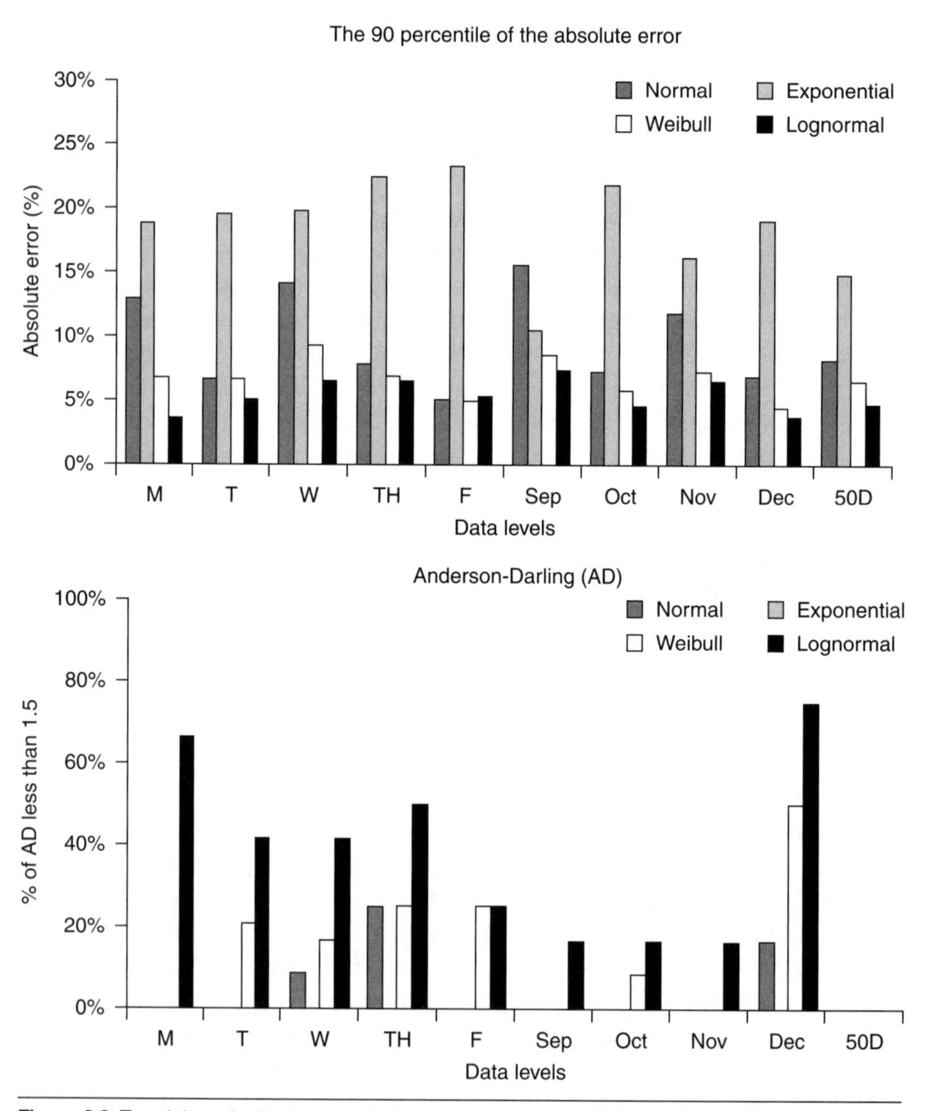

Figure 3.3 Travel time distributions evaluation criteria

example, the correlation of travel times on segments S_{11} and S_{12} is obvious with a correlation coefficient of 0.70 and *p*-value less than 0.0001 (see Table 3.3 for Mondays, 2003). Table 3.3 indicates that the correlation coefficients between travel times on these two adjacent segments are significant. However, for the rest of the segments that are adjacent to each other (S_{12} and S_{13}, S_{13} and S_{14}, S_{14} and S_{15}), correlation is insignificant. Note that although the correlation coefficient between S_{12} and S_{13} is

Table 3.3 Correlation between I-4 eastbound segments on Mondays in 2003

Pearson Correlation Coefficients, N = 480
Prob > |r| under H_0: $\rho = 0$

Segment	S_{11}	S_{12}	S_{13}	S_{14}	S_{15}
S_{11}	1	0.70058 <.0001			
S_{12}	0.70058 <.0001	1	0.16795 0.0002		
S_{13}		0.16795 0.0002	1	0.12072 0.0081	
S_{14}			0.12072 0.0081	1	0.05605 0.2203
S_{15}				0.05605 0.2203	1

significant, it is weak (only 0.16795). The analysis was based on computing and testing correlation coefficients for each pair of segments, using the data points available for each pair. The Pearson correlation coefficient (product-moment coefficient of correlation) was used in this analysis. It measures the strength of a *linear* relationship between a pair of two variables:

r: Pearson Correlation Coefficient (Sample)
ρ: Population Correlation Coefficient

The result is that for these travelers who traveled segments S_{11} and S_{12}, it would be more accurate to calculate the travel time mean and variance for segment S_{21}, that starts at loop detector ID_5 (on ramp, E. US 192) and ends at loop detector ID_22 (off ramp, 75A-Universal Drive) as shown in Table 3.1, instead of summing the mean and variance of segments (S_{11} and S_{12}) without consideration of their travel time correlation. Note that segment S_{21} includes segments S_{11} and S_{12} as part of its length.

3.5.3 Travel Time Reliability Stochastic Models

The reliability function $R(t)$ of a system at time t is the cumulative probability that the system has not failed from 0 to t. This can be represented as:

$$R(t)_n = P\{T_i \geq t\} = 1 - \int_0^t f(t)dt = 1 - F(t), -\infty < t < \infty \tag{3.9}$$

Accordingly, the best-fit (lognormal) travel time reliability model in Equation (3.6) was compared to the empirical cumulative distribution reliability function (ECDRF). The ECDRF supplies immediate information regarding the shape of the underlying distribution, outliers, and robust information on location and dispersion. The ECDRF of a sample $\{T_i, i = 1, 2, 3, ..., n\}$ is defined as the following function:

$$R_n = 1 - F_n = 1 - \frac{\#(T_i \leq t)}{n}, \; -\infty < t < \infty \tag{3.10}$$

Where # $(T_i \leq t)$ is read as the number of (T_is) less than or equal to (t) for the sample size (n).

3.5.4 Theoretical Versus Empirical Reliability Distributions

Figure 3.4 shows plots of 3 out of the 12 analyzed segments to examine the convergence between ECDRF and the lognormal stochastic models and the similarities among the weekdays (that is Mondays, Thursdays, and Fridays) for the same segment. Clearly, the two plots (ECDRF and the lognormal stochastic models) are almost identical for the same weekday, which means that there are no significant differences between them.

After the visual inspection, the t-statistical test was used to determine if the two means of the empirical and the best-fit lognormal stochastic models are equal assuming equal variances. The results showed no significant difference at 95

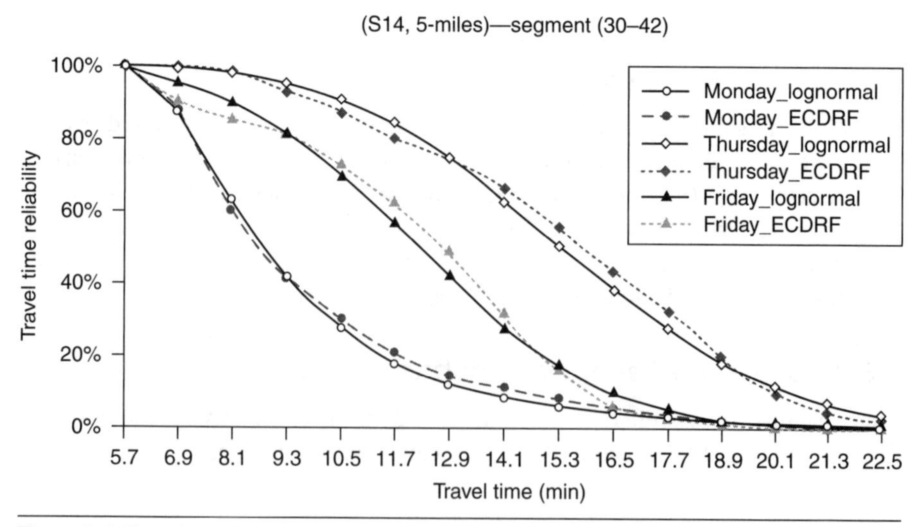

Figure 3.4 The relationship between empirical cumulative distribution reliability and best-fit distributions

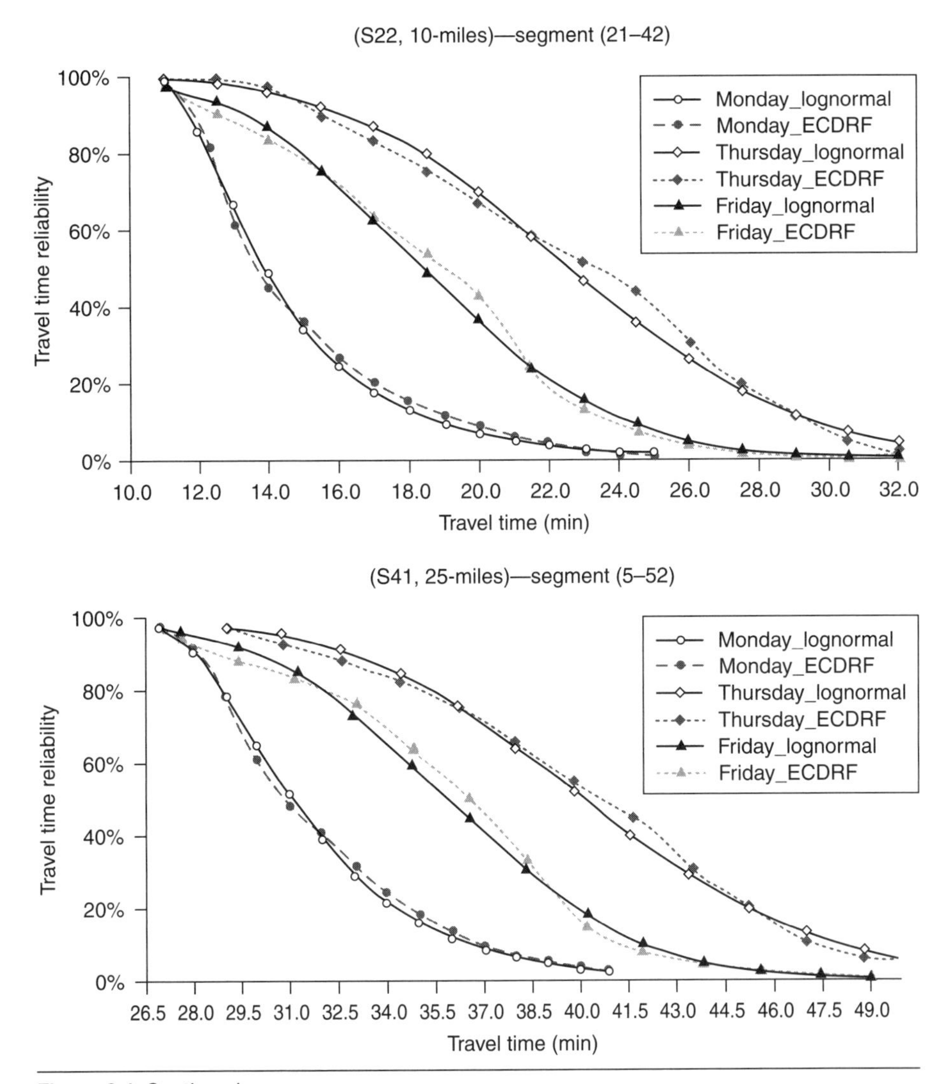

Figure 3.4 *Continued*

percent confidence level between the mean of the two distributions (lognormal and ECDRF) for all segments (that is, $P > 0.05$). It can be concluded that the lognormal models replicate the ECDRF reliability distributions at the 95 percent confidence level.

Furthermore, the t-statistical test was used to see if there was a statistically significant difference between the weekdays which would require different parameter estimates (models) for each weekday data group (that is, Mondays, Tuesdays,..., and Fridays). Using the t-student test (two samples with unequal variances) after taking the logarithm of the travel time, because the logarithm of a lognormal is normal, it was found that there was a significant difference (that is, $P < 0.05$) between the weekdays (Mondays through Fridays) in most of the segments (for example, S_{13}, S_{22}, and S_{41}). Consequently, it was imperative to compute the appropriate estimates of the lognormal model parameters based on the day of the week for each segment.

3.5.5 Segment Travel Time Reliability Models

Unlike mechanical equipment, it is unrealistic that any freeway segment could be traversed in zero seconds no matter how fast the vehicles are; therefore, an adjustment of the model location parameter was needed to accurately estimate the travel time reliability and to achieve the following definition: A roadway segment is considered 100 percent reliable if its travel time is less than or equal to the travel time at the posted speed limit. The threshold or the location parameter γ had to be adjusted to reflect the acceptable travel time level (travel time at posted speed limit). The adjusted location parameter can be calculated as a function of the acceptable upper limit of travel time and the corresponding reliability as follows:

$$\sigma\,\Phi^{-1}(1-R(T)) = Ln(t-\gamma) - Ln(T_{50}) \;\Rightarrow\; Ln(t-\gamma) = Ln(T_{50}) + \sigma\,\Phi^{-1}(1-R(T))$$

Substituting for t = distance divided by speed limit and considering that $R(T)$ is 100 percent, we get:

$$Ln(t-\gamma) = Ln(T_{50}) - 3\sigma \;\Rightarrow\; \gamma = t - e^{[Ln(T_{50})-3\sigma]}$$

The reason for substituting $\phi^{-1}[(1 - R(T))] \approx -3$ is that *under a normal curve, approximately 99 percent of test observations fall within three standard deviations.* Similarly, the adjusted location parameter γ can be calculated for the other three models (Weibull, exponential, and normal).

In Figures 3.5 and 3.6, it is observed that there are large variations between weekdays in travel time reliability. Since the travel times had different patterns among the weekdays, different models for each weekday were developed for each segment.

3.5.6 Travel Time Reliability and Departure Time

The developed lognormal models were used to estimate the travel time reliability as a function of the departure time for each weekday for the 12 analyzed segments as described in the model application section. The uncongested segments' (for example, S_{21}) travel time reliability of the weekdays had the same pattern and varied by no more than 20 percent among the weekdays as shown in Figure 3.5. This depicted the sensitivity of the model for any minor travel time changes in such segments and resulted in continuous fluctuation in the travel time reliability pattern with the departure time. At the end of the peak, this fluctuation decreased to around 10 percent. On the other hand, the congested segments (for example, S_{14} and S_{31}) had an obvious peak (starting between 17:15 and 17:30) in which the travel time reliability dropped to less than 40 percent, as shown in Figure 3.5.

As expected, the longer the segment, the higher the variation in the travel time reliability among the weekdays. For example, there is a noticeable difference in the

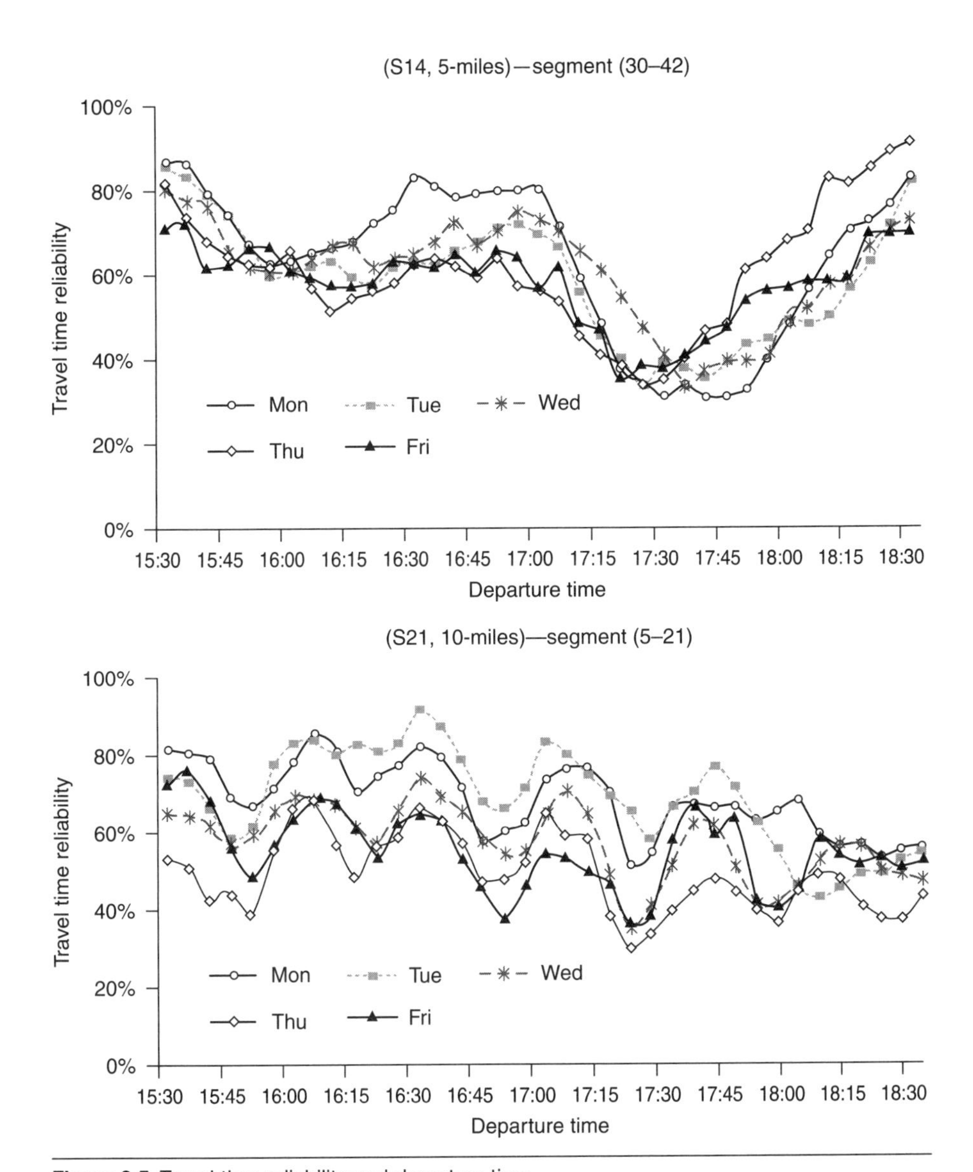

Figure 3.5 Travel time reliability and departure time

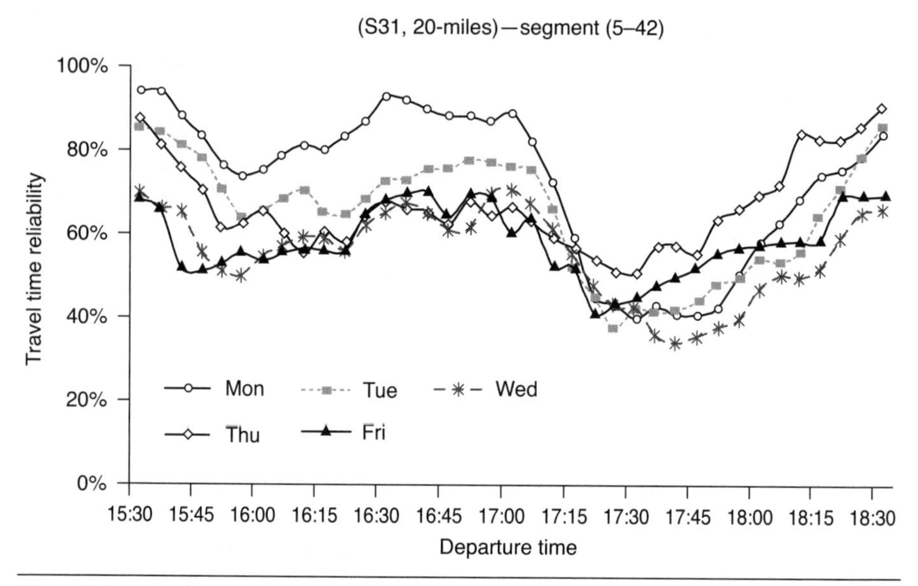

Figure 3.5 *Continued*

fluctuation of the travel time reliabilities prior to 17:00 for segment S_{14} compared to S_{31} among the weekdays at the same departure time as shown in Figure 3.5. In this case, segment S_{31} is about 20 miles and is composed of uncongested and congested segments (S_{21}, S_{13}, and S_{14}), while segment S_{14} is only 5 miles.

3.5.7 Benefits of the New Method to Practitioners and Travelers

The results underscore the fact that the developed models are appropriate for evaluating freeway operation performance. They capture the travel time variability experienced by individual travelers over an extended period of time and can be used to calculate and predict travel time reliability of the freeway segment or corridor in real time or to assess historical performance. The following demonstrates the benefits of the new method and compare it with existing methods.

3.5.8 Segment Travel Time Reliability

Figure 3.6 illustrates how travel time reliability of a segment (using the new methodology) varied significantly with the day of the week, segment location, and departure time. It is obvious that, for some segments, the travel time reliability at 18:30 (end of peak) is higher than the corresponding values at 17:30 (peak). For example, segment S_{14} travel time reliability ranges from 31 to 41 percent at 17:30

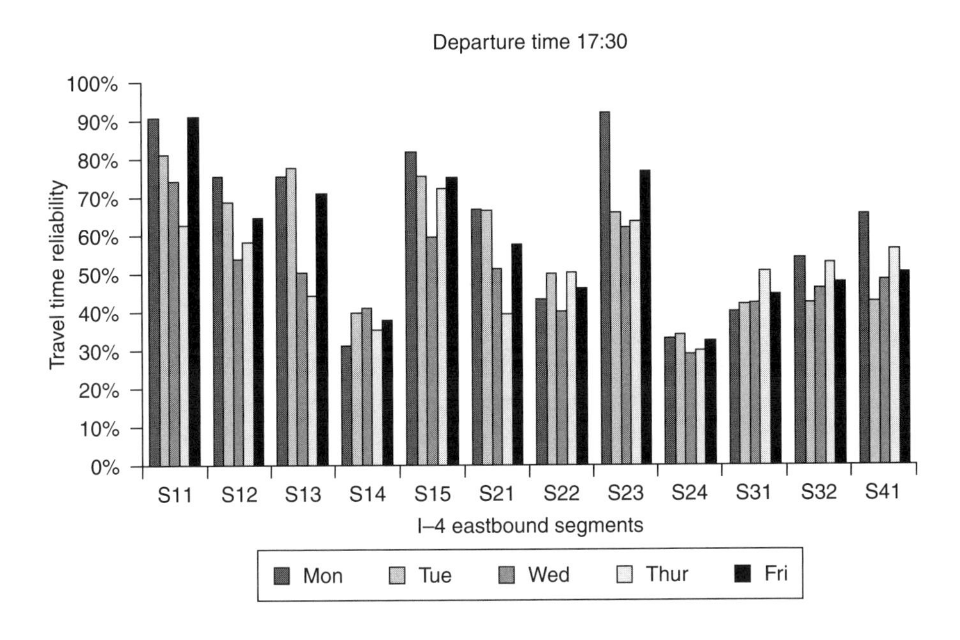

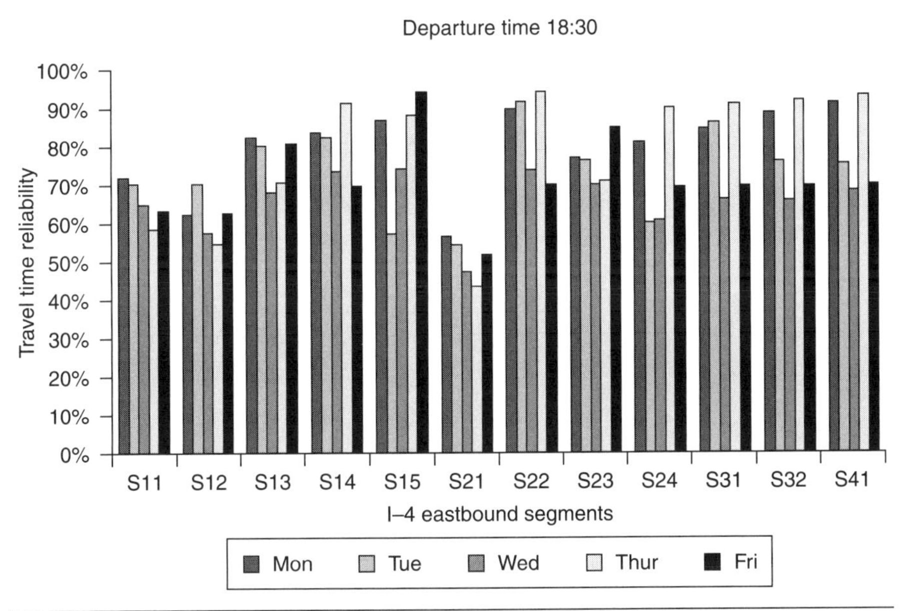

Figure 3.6 Travel time reliability and departure time for all segments

compared with 70 to 91 percent at 18:30 on weekdays. This information can be disseminated to roadway users during their trip planning to help them make a decision about their departure time. The travel time reliability can also be disseminated in-trip through cell phones, radio, or changeable message signs (CMS) to eliminate freeway user anxiety about the unknown, guide them to the diversion alternatives, and give them the probable time they will reach their destinations.

3.5.9 Corridor Travel Time Reliability

Figure 3.7 illustrates and points to the strength of the new methodology because it captures the freeway performance by treating travel time as a continuous variable. The corridor reliability was estimated using the series system as in Equation (3.9). The corridor reliability is equal to the product of the reliabilities of its components. This means that the segment with the smallest reliability has the greatest effect on that corridor's reliability. The travel time reliability of the corridor as a function of the departure time has the same trend, but it significantly depends on the segment length. The smaller the segment, the lower the corridor travel-time reliability and the more accurate the estimated reliability is as shown in Figure 3.7.

It was not surprising to see low corridor reliabilities throughout the evening peak period with the 5-mile segments compared to the 25-mile segment. For the 5-mile segments, the freeway users judge each segment individually and any increase in travel time, particularly on the uncongested segments, has a significant

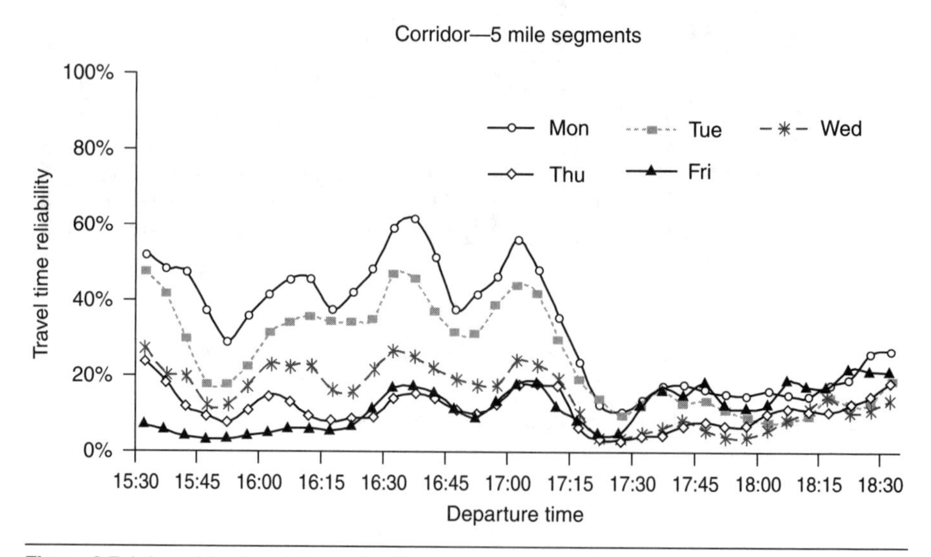

Figure 3.7 I-4 corridor travel time reliability versus departure time

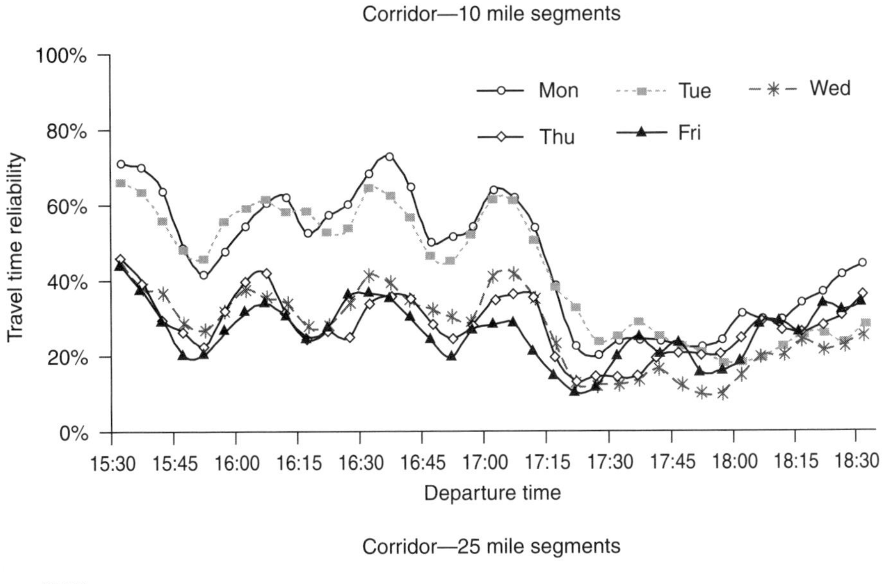

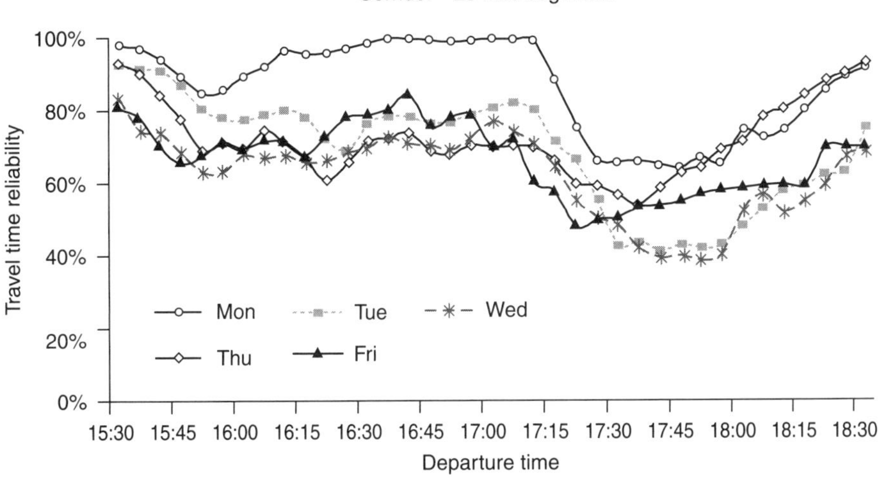

Figure 3.7 *Continued*

effect on the overall corridor travel time reliability. Accordingly, the corridor travel time reliability was found to be less than 20 percent during the peak period from 17:15 to 18:15. This information is more appropriate for freeway operations as a microscopic performance measure of the freeway corridor. On the other hand, the corridor travel time reliability was found to be at least 40 percent during the same period using the 25-mile segment as shown in Figure 3.7. As such, using long segments in calculating travel time reliability can result in misleading information.

Long segments (that is, 25 miles) may show that the corridor travel time is reliable, while short segments (that is, 5 miles) may indicate that the corridor is unreliable due to severe congestion. However, using long segments may be more appropriate for planning purposes as a macroscopic evaluation of the freeway corridor, especially in the absence of detailed data for smaller segments.

Evidently, the travel time reliability is strongly dependent on the length of the analyzed segments and is highly affected by the reliability of its bottlenecks and sections congested upstream of the bottlenecks. We were able to show the high discrepancy in the corridor travel time reliability due to high sensitivity to segment length. Therefore, it is recommended that one should calculate the travel time reliability for smaller segments and then calculate the corridor reliability using the series equation for performance evaluation purposes.

3.5.10 Comparison Between Reliability Methods

This section compares the newly-developed method and the existing two methods (Florida and BT methods) that are ranked highly in the literature.

Table 3.4 illustrates application of the BT reliability method (as defined in the literature review) using the 4-month data sample of the I-4 evening commute.

Table 3.4 Example application of buffer time reliability measure (minutes) using a data sample from the I-4 data warehouse

		Data levels (%)				
Segment	Distance	M	T	W	TH	F
S11: 5–14	5.71	0.35	0.35	0.93	0.34	0.37
S12: 11–22	6.38	0.27	0.23	0.77	0.41	0.29
S13: 21–30	4.95	0.34	3.52	3.95	2.75	3.43
S14: 30–42	5.19	4.76	5.87	5.36	5.26	6.83
S15: 42–52	4.83	3.02	3.76	3.78	3.41	1.65
S21: 5–22	10.14	0.47	0.43	1.38	0.57	0.56
S22: 21–42	10.14	7.21	5.60	8.67	7.48	6.95
S23: 11–30	10.84	0.66	3.68	4.51	2.70	3.40
S24: 30–52	10.02	7.26	6.52	7.17	7.11	5.15
S31: 5–42	19.79	7.23	5.59	9.58	7.28	6.46
S32: 11–52	20.86	7.68	7.09	10.30	10.03	6.66
S41: 5–52	24.62	7.45	7.20	10.54	10.19	6.57
Corridor buffer time	(5 miles)	8.73	13.73	14.78	12.17	12.57
	(10 miles)	10.70	9.78	13.83	11.46	9.16
	(25 miles)	7.45	7.20	10.54	10.19	6.57

The BT was calculated as follows: BT= (95th percentile travel time-average travel time). For example, the average weekday corridor (Tuesday) BT needed to ensure on-time arrival for 95 percent of the trips was 13.73 minutes (based on the 5-mile segments). This means that a traveler should budget an additional 38.6 percent buffer for a 35.59 minute average peak trip time for the 25-miles on the I-4 corridor to ensure a 95 percent probability of on-time arrival. Using the same data sample, Table 3.5 demonstrates application of the Florida reliability method with 5 percent excess travel time over the median (34.56 minutes). A statewide (or maybe a national) traveler survey is needed to determine the appropriate percentage. In this Florida method, Shaw and Jackson (2003) recommended the Weibull distribution for the calculation of the reliability statistic as follows:

$$R(t) = P(x < X + \Delta) = P(x < TT) = 1 - e^{-(TT/\eta)^{\beta}}$$

where β is defined as the shape parameter and η is the scale parameter.

The other parameters of this equation were defined in the literature review. The following is a numerical example to illustrate the computations in Table 3.5. The average weekday (Tuesday) percent of travel that takes no longer than the

Table 3.5 Example application of the FDOT reliability measure (%) using a data sample from the I-4 data warehouse

Segment	Distance	Data levels (%)				
		M	T	W	R	F
S11: 5–14	5.71	90.82	91.70	73.77	89.10	89.50
S12: 11–22	6.38	94.00	97.84	78.25	86.10	95.47
S13: 21–30	4.95	74.74	52.36	39.22	66.17	43.17
S14: 30–42	5.19	88.72	57.12	54.61	60.29	15.77
S15: 42–52	4.83	56.46	51.16	52.80	52.68	62.26
S21: 5–22	10.14	95.08	97.57	76.45	89.60	94.22
S22: 21–42	10.14	53.53	59.38	52.23	63.96	63.32
S23: 11–30	10.84	81.76	68.08	51.60	69.28	57.49
S24: 30–52	10.02	52.94	58.54	53.61	61.09	63.39
S31: 5–42	19.79	59.15	65.67	55.10	63.94	67.77
S32: 11–52	20.86	58.59	63.00	54.43	63.59	68.15
S41: 5–52	5.83	60.17	64.62	55.55	64.47	69.74
Corridor reliability	(5 miles)	31.97	13.73	6.53	16.13	3.62
	(10 miles)	28.74	29.64	21.08	30.19	37.14
	(25 miles)	60.17	64.62	55.55	64.47	69.74

Note: This table uses 5 percent above median travel time as one of the tolerance thresholds proposed by FDOT (see Shaw and Jackson, 2003)

median travel time to traverse the 25 miles of the I-4 corridor (34.56 minutes) plus 5 percent tolerance for additional time (TT = 34.56 × 1.05 = 36.31 minutes) had a value of 13.73 percent. In other words, 13.73 percent of the freeway commuters travel the 25-mile segment in less than 36.31 minutes.

Similar to the newly-developed method, the corridor travel time reliability (25 miles) calculated in both methods using the 5-mile segments are more accurate than using the full length of the 25 miles as one long segment. Again, the smaller the segment is, the more accurate the corridor travel-time reliability is.

Figure 3.8 shows a graphical comparison between the three methods when applied to a 5-mile segment (S_{15}) eastbound I-4 for a Thursday 2003 data sample. This segment had a 5.3 minute travel time at the speed limit as shown in Figure 3.8a. According to the new method, this segment has at least 95 percent travel time reliability as long as its travel time does not increase by more than 1.53 minutes (or 28.87 percent of its free-flow travel time). The basic premise of this new method is that it is sensitive to the users' perspective because it reflects that an increase in segment travel time should always result in less travel time reliability. To illustrate further, if travel time is doubled (from 5.3 minutes to 11.0 minutes), then reliability would have decreased down to only 20.1 percent as shown in Figure 3.8a.

Figure 3.8b illustrates application of the FDOT method. The median travel time for segment S_{15} was 7.5 minutes, and the corresponding travel time reliability for the suggested percentages of 5, 10, 15, and 20 percent were: 57.6, 64.4, 70.6, and 76.2 percent respectively. According to the Florida method, for segment S_{15} to achieve 95 percent reliability at a travel time of 11.0 minutes, the users would

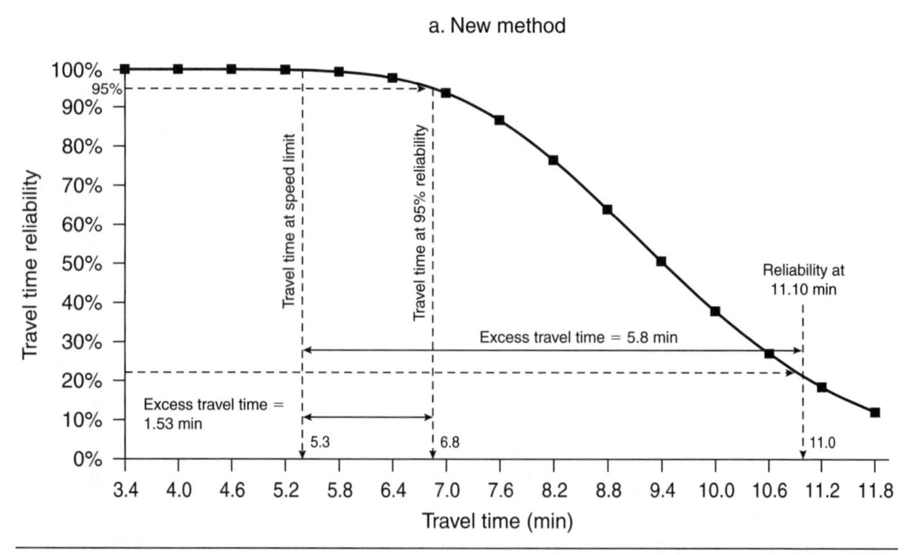

Figure 3.8 Comparison of the reliability methods for segment S_{15} (5-mile) for Thursdays, 2003

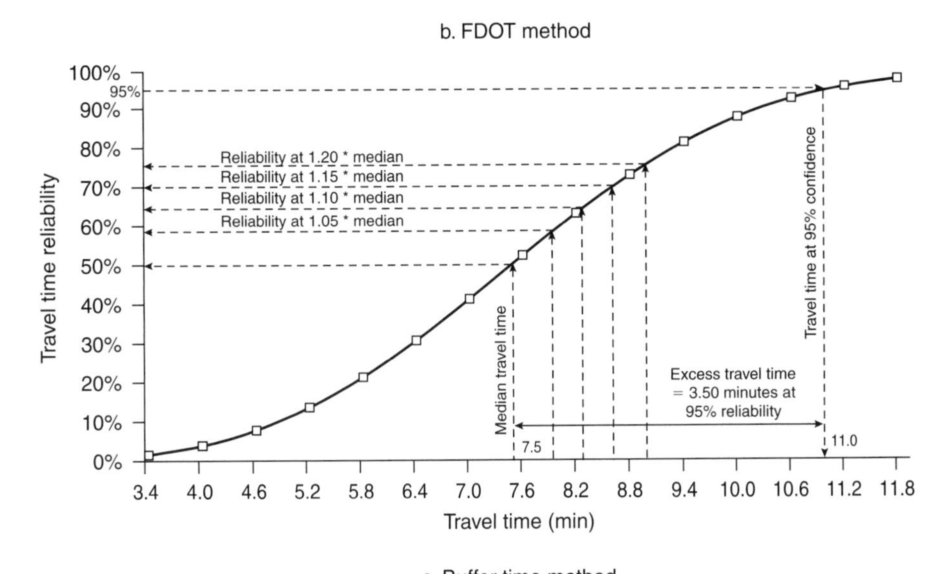

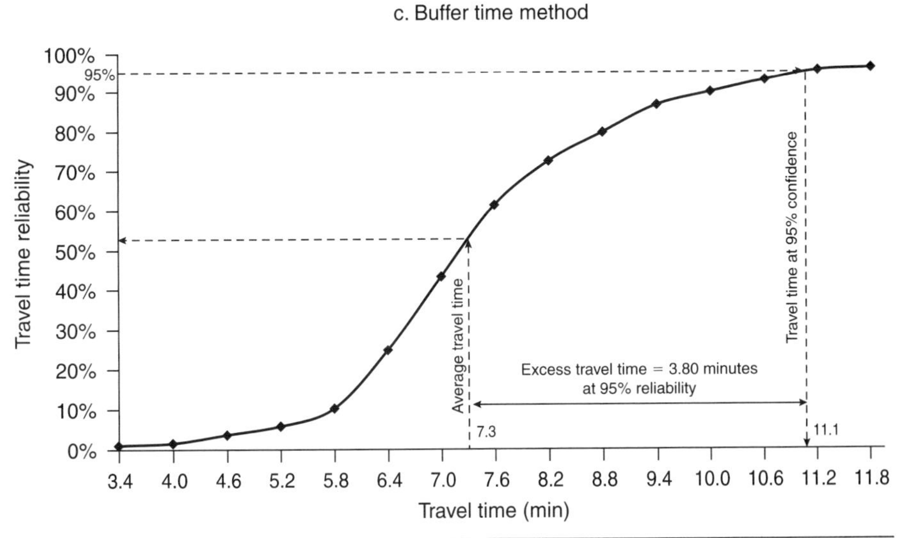

Figure 3.8 *Continued*

have to accept an increase of 46.7 percent in excess travel time above the median (or an excess of 3.5 minutes above 7.5 minutes). In contrast, the new method's estimated reliability for this segment is only 20.1 percent for the same travel time of 11.0 minutes.

Figure 3.8c illustrates application of the BT method to segment S_{15}, which had an average travel time of 7.3 minutes. The BT, or excess time above the average, for this segment was 3.80 minutes or about 50 percent in excess travel time above the average. So, according to the BT method, travel time needed to ensure that 95 percent of the users are on time is 11.1 minutes. This is close to 11.0 minutes, which is the same travel time for which the Florida method estimated reliability at 95 percent for this segment. This indicates that both BT and FDOT methods performed similarly. At least a 47 percent tolerance above the mean or the median is required to achieve the 95 percent on-time arrival. Similar results were obtained from the 10-mile segment (for example, S_{22}) and the corridor analysis (25-mile segment).

To illustrate the new method further, a comparison between the three methods was applied to a 10-mile segment (S_{22}) with a 12.2 minute travel time at speed limit. This segment has at least 95 percent travel time reliability as long as its travel time does not increase by more than 6.33 minutes (new method). Travel time reliability decreased to 24.7 percent for an increase of 18.2 minutes (from 12.2 minutes to 30.4 minutes). The FDOT and BT methods travel time median and mean were 22.7 minutes and 23.2 minutes respectively. The travel time needed to ensure that 95 percent of the users are on time is 30.4 minutes using the BT method, which is close to 31.2 minutes (same travel time for which the Florida method estimated reliability at 95 percent for this segment). This indicates that both BT and FDOT methods performed similarly and at least a 34 percent excess in the tolerance above the mean or the median is required to achieve 95 percent arrival on time.

The corridor analysis (25-mile segment) results were also calculated. The new method shows the corridor has at least 95 percent travel time reliability as long as its travel time does not increase by more than 6.6 minutes or 22.38 percent above the travel time at speed limit (29.5 minutes). Travel time reliability decreased to 18.4 percent for an increase of 20.9 minutes. Both the FDOT and BT methods have similar results in which an increase of about 25 percent of the mean or the median is required to achieve 95 percent on-time arrival.

It is interesting to see that both Florida and BT methods are close in their reliability estimates. Clearly, there is inconsistency between the reliability estimates of the new method and the existing methods used in this numerical example. This finding is not counterintuitive because the key difference between the new method and the existing methods is that the new method's definition of reliability is totally different from the existing methods. The developed method is the only method that did not deal with the cumulative travel time distribution only, but rather with the reliability distribution itself which considers the users' perspective in the sense that an increase in travel time means less reliability of the corridor or the segment. Using warehouse data, we are able to illustrate the vulnerability of the existing reliability methods where the traveler was expected to accept 95% reliability while experiencing a congested segment with travel time more than twice the free-flow travel time. The ultimate judge of which method should prevail is the traveler. As such, traveler surveys are critical to make this determination.

3.6 SUMMARY

A new methodology for estimating travel time reliability in a freeway corridor was developed in this chapter by using a traffic data warehouse. A four-month data sample was studied. The methodology was applied to a section of I-4 in Orlando, Florida. Four different stochastic models were tested: Weibull, exponential, lognormal, and normal. Two evaluation criteria were used in selecting the best-fit model: (1) Anderson-Darling (AD) GoF statistic and (2) 95th error percentages. Based on these criteria, the lognormal model provided the best fit of travel time reliability for the evening peak period of the I-4 corridor (eastbound). It was also shown to be more efficient to use the same day of the week for estimating travel time reliability for I-4 segments rather than using mixed data such as an entire month because of the significant differences among weekdays in the same month.

The developed lognormal model was used to estimate segment and corridor travel time reliability based on the following definition: A roadway segment is considered 100 percent reliable if its travel time is less than or equal to the travel time at the posted speed limit. This definition puts more emphasis on the users' perspective and, as such, is different from reliability definitions of the existing methods of FDOT and BT. The vulnerability of the existing methods was demonstrated using warehouse data compared with the new method.

The developed travel time reliability model was dependent on the length of the analyzed segments. If there is a lack of detailed travel time data of smaller segments, longer segments can be used to estimate freeway reliability, but this is only appropriate for planning purposes. On the other hand, if detailed data are available, then smaller segments should be used to evaluate travel time reliability of freeway corridors for the purposes of freeway operations. Furthermore, the new method showed high sensitivity to the geographical location which reflects the level of congestion and bottlenecks. A major advantage of the new method above the existing ones is its strong potential ability to estimate travel time reliability as a function of departure time. The new method is more appropriate for freeway operations because it treats travel time as a continuous variable that captures the variability experienced by individual travelers over an extended period.

The newly-developed method has a strong potential for computing and predicting travel time reliability of the freeway corridor in real time and assessing the historical performance of freeway corridors. Future research will focus on traveler surveys to determine the travelers' perception of travel time reliability and acceptable thresholds of delay above free-flow travel time. We also plan to apply this new methodology to peak and off-peak periods (for example, morning and midday) using the I-4 traffic data warehouse. Finally, the methodology used in this study may be transferable to other similar freeway surveillance data for derivation of customer service. The new method can be extended to compute reliability for large-scale real-life networks.

REFERENCES

Al-Deek, H. and R. Chandra. 2004. New algorithms for filtering and imputation of real time and archived dual-loop detector data in the I-4 data warehouse. *Journal of the Transportation Research Board*, 1867. (January 1995). 116–126.

Al-Deek, H., P. Kerr, B. Ramachandran, J. Pooley, A. Chehab, E. Emam, R. Chandra, Y. Zuo, K. Petty, and I. Swinson. 2004. The Central Florida Data Warehouse (CFDW), Phase-2-, Final Research Report, Transportation Systems Institute (TSI), Department of Civil & Environmental Engineering, Orlando: University of Central Florida.

Al-Deek, H. and E. Emam. 2006. New methodology for estimating reliability in transportation networks with degraded link capacities. *Journal of Intelligent Transportation Systems*. In press.

Al-Fawzan, M. 2000. Algorithms for estimating the parameters of Weibull distribution. King Abdulaziz City for Science and Technology, Saudi Arabia, (October 20). Available from http://ip.statjournals.net:2002/InterStat/ ARTICLES/2000/articles/O00001.pdf.

Asakura, Y. and M. Kashiwadani. 1991. Road network reliability caused by daily fluctuation of traffic flow. Proceedings of the 19th PTRC Annual Meeting, Brighton, 73–84.

Bell, M. and C. Cassir. 2000. *Reliability of Transport Networks*. Baldock, UK: Research Studies Press.

Bell, M. and Y. Iida. 2003. *The Network Reliability of Transport*. Oxford, UK: Pergamon-Elsevier Science.

Chen, C., A. Skabardonis, and P. Varaiya. 2003. Travel time reliability as a measure of service. *Journal of the Transportation Research Board*, 1885: 74–79.

Chen, A. and W. Recker. 2000. Considering risk taking behavior in travel time reliability. UCI-ITS-WP-00-24, Institute of Transportation Studies. Irvine: University of California. Available: http://www.its.uci.edu.

Chen, A., H. Yang, H. K. Lo, and W. Tang. 1999. A capacity-related reliability for transportation networks. *Journal of Advanced Transportation*, 33: 183–200.

D'Agostino, R. and M. Stephens. 1986. *Goodness-of-Fit Techniques*. New York: Marcel Dekker.

FDOT. 2000. The Florida reliability method: In Florida's mobility performance measures program. http://www.dot.state.fl.us/planning/statistics/mobility-measures (accessed March 31, 2004).

Kane, V. E. 1982. Standard and goodness-of-fit parameter estimation methods for three parameter lognormal distribution. *Communications in Statistics*, 11: 1935–1957.

Kececioglu D. 1991. *Reliability Engineering Handout, Volume 1*. Upper Saddle River, NJ: Prentice Hall.

Levinson, D. and L. Zhang. 2001. Travel time variability after a shock: The case of the Twin Cities ramp meter shut off. Presented at the 1st International Symposium on Transportation Network Reliability, Kyoto, Japan.

Lomax, T., S. Turner, M. Hallenbeck, C. Boon, R. Margiotta, and A. O'Brien. Traffic congestion and travel reliability: How bad is the situation, and what is being done about it? U.S. Department of Transportation, Federal Highway Administration. (accessed May 20, 2004) http://www.fhwa.dot.gov/congestion/cgstpapr.htm.

Mendenhall, W. and T. Sincich. 1995. *Statistics for Engineering and the Sciences*, 4th ed. Upper Saddle River, NJ: Prentice Hall.

NCHRP Synthesis 311, National Cooperative Highway Research Program. 2003. Performance measures of operational effectiveness for highway segments and systems. A synthesis of highway practice, Consultant: Terrell Shaw and PBS&J, Transportation Research Board.

Rencher, A. 2002. *Methods for Multivariate Analysis*, 2nd ed. John Wiley & Sons.

Shaw, T. and D. Jackson. 2003. "Reliability performance measures for highway systems and segments." Paper presented at the 82nd Transportation Research Board Annual Meeting, Washington, DC.

Shimokawa, T. and M. Liao. 1999. Goodness-of-fit tests for Type-I extreme-value and 2-parameter Weibull distributions. *IEEE Transactions on Reliability*, 48(1): 79–86.

Tobias, A. and C. Trindade. 1995. *Applied Reliability*, 2nd ed. New York: Van Nostrand Reinhold.

Turner, S., M. Best, and D. Schrank. 1996. Measures of effectiveness for major investment studies. Southwest Region University Transportation Center Report SWUTC/96/467106-1, Texas Transportation Institute. College Station, TX: Texas A&M University.

Yang, H., H. Lo, and W. Tang. 2000. *Travel Time versus Capacity Reliability of a Road Network*. N. Bell, C. Cassir, eds. Reliability of Transport Networks, Research Studies Press.

William L. and R. Laurence. 2002. Examining information needs for efficient motor carrier transportation by investigating travel time characteristics and logistics. Texas Transportation Institute Report No. SWUTC/01/473700-00005-1.

4

MIXED LOGIT MODELING OF PARKING-TYPE CHOICE BEHAVIOR

Stephane Hess
Institute for Transport Studies
University of Leeds

John W. Polak
Centre for Transport Studies
Imperial College–London

4.1 INTRODUCTION

Parking policy is an important component of contemporary travel-demand management policies. The effectiveness of many parking policy measures depends on influencing parking-type choice, so that understanding the factors affecting these choices is of considerable practical importance. Yet, academic interest in this issue has been intermittent at best. This chapter reports the results of an analysis of parking-choice behavior, based on a stated choice (SC) data set, collected in various city center locations in the United Kingdom. The analysis advances the state of the art in the analysis of parking-choice behavior by using a mixed multinomial logit (MMNL) model, capable of accommodating random heterogeneity in travelers' tastes and the potential correlation structure induced by repeated observations made of the same individuals. The results of the analysis indicate that taste heterogeneity is a major factor in parking-type choice. Accommodating this heterogeneity leads to significantly different conclusions regarding the influence of substantive factors such as access, search and egress time, and the treatment of potential fines for illegal parking. It also has important effects on the implied willingness to pay for time-savings and on the distribution of this willingness across the population. Our analysis also reveals important differences in parking behavior across different journey purposes, as well as across the three locations used in the SC surveys. Finally, the chapter discusses a number of technical issues related to the specification of taste heterogeneity that are of wider significance in the application of the MMNL model.

The development of an efficient parking policy is an important component of urban transport planning, as it can help ease congestion and improve the competitiveness of city centers as well as the quality of life in residential areas. The objective of many parking policy measures is to influence choices made during the parking process, therefore, it is important to gain an understanding of the factors affecting parking behavior.

One important aspect of the study of travelers' parking behavior is the modeling of the choice of parking type. In this chapter, we conduct a disaggregate modeling analysis of this choice process, using an SC data set collected in three locations in the United Kingdom in 1989. It should be expected that, as in many other areas of choice-making behavior, there are important differences among decision makers in their reactions to changes in various attributes of a given alternative, in this case a specific type of parking. Such differences in choice-making behavior are known as taste variation, and their existence signals a departure from a purely homogeneous population of decision makers. While it is possible to explain some part of this variation in a deterministic way, for example by a segmentation of the population into mutually exclusive subsets or by linking tastes continuously to sociodemographic indicators, there is, in many situations, an additional, purely random variation in tastes within groups of decision makers (as opposed to variation between groups). So far, the literature on parking behavior has largely ignored the possibility of such a random variation in tastes. Given the potential bias that can be caused by ignoring such variation, in addition to the poorer model fit, this omission seems to be a major gap in the present area of research. The aim of our analysis is thus to test for the presence of random, as well as deterministic taste variation, and to quantify the impacts of incorporating such random variations in tastes in terms of coefficient values as well as model performance. For this, we employ an MMNL model that allows for both types of taste heterogeneity and which, furthermore, explicitly accounts for the repeated choice nature of the data set. In our attempts to explain the deterministic differences in choice behavior across locations and journey purposes, the data set is split into subsets and a separate model is estimated for each subset.

The remainder of the chapter is organized as follows: the second section gives a brief review of existing studies of parking choice; the third describes the data used in the present analysis, and the fourth describes the modeling approach used. Section five presents and discusses the findings of the modeling analysis for the different data sets used, and the sixth section presents our conclusions

4.2 REVIEW OF PARKING-TYPE CHOICE LITERATURE

The SC data used in the present analysis has previously been used in the modeling of choice of parking type by Axhausen and Polak (1991). As part of their research, Axhausen and Polak reviewed previous studies of the choice of parking type, noting that, while the effects of parking costs and times on mode choice had been investigated in great detail, few applications had, up to that point, investigated the choice between different types of parking; a situation that has not changed greatly.

Research that is of special importance in the current analysis is that based on the use of discrete choice models; a brief overview of such studies appears in Section 4.4.2. In terms of the actual parking-type choice process modeled in the present paper, only a small number of studies are important. One example is the application given by Van der Goot (1982) who groups the parking options not just by location, but also by type of parking. Another example is given by Hunt (1988) who conducts an analysis similar to ours, not just in terms of using logit type models, but also in terms of the types of parking considered. Aside from these two studies, most existing research has looked at more general transport issues that have some parking component. For a more detailed review of such studies, see Axhausen and Polak (1991) or Polak and Vythoulkas (1993).

4.3 DATA

4.3.1 Description of Data and the Stated Choice Survey

The SC data set used in this chapter was collected for an analysis of parking behavior in the West Midlands region of the United Kingdom (cf. Polak et al., 1990). SC surveys were conducted in 1989 in the central business districts of Birmingham and Coventry and in a suburban center, Sutton Coldfield. Although the data are now somewhat dated, the main aim of the chapter is to explore the relevance of simulation-based random coefficients models in the modeling of parking-type choice (rather than to undertake policy analysis per se), so this is not regarded as a significant drawback.

Respondents were selected at street level on the basis of certain screening criteria and target quotas—concerned with socio-economic as well as journey-related factors (see Polak and Axhausen, 1989). Five types of parking options were considered:

1. Free on-street parking
2. Charged on-street parking
3. Charged off-street parking
4. Multistory car parking
5. Illegal parking

The types of parking were described by a set of four attributes: access time (to the parking area), search time (for a parking space), egress time (walking time to final destination), and cost. The cost attribute was set to zero for the free on-street alternative and was replaced by the expected fine for the illegal parking alternative. The expected fine was calculated by multiplying the probability of receiving a ticket with the level of fine currently in use (the probability of the fine actually being enforced was treated as an unknown factor).

Respondents were asked to provide details about their current type of parking along with two possible alternative parking options for their current journey, so that the set of three alternatives comprised illegal parking and two different legal parking options. Using this information, along with the attributes for the three alternatives, 81 choice situations were constructed, each including the three given parking options but with the attributes of the alternatives being varied according to an orthogonal SC design (cf. Polak and Axhausen, 1989). Four journey purposes, or activities, were identified in the SC survey; these were full-time and part-time work trips, shopping trips, and errand trips. In the ensuing SC experiment, each respondent was presented with a fractional factorial block of SC situations (for the original journey-type used) drawn at random from the 81 possible choice situations. For more details on the actual SC survey, see Polak and Axhausen (1989).

In the present analysis, two divisions of the data set were used, grouping respondents by location (three groups) as well as by activity. Because of the low number of part-time work and errand trips, the activity division defined only two groups: work trips (both full-time and part-time) and shopping and errand trips. The resultant distribution of attribute values across alternatives is summarized in Table 4.1, giving the minimum, mean, and maximum values for each attribute in the various groups. Data from a total of 1,335 choice situations were used, collected from a sample of 298 respondents. The data are summarized in Table 4.2, giving the number of observations by location and activity as well as the number of times that each alternative was included in the SC situations.

Table 4.1 Attributes of parking options across locations and journey purposes

	Birmingham			Sutton Coldfield			Coventry			Work (FT & PT)			Shopping and errands		
	min	mean	max	min	mean	max	min	mean	max	min	mean	max	min	mean	max
Free on-street															
Access time	1	23.3	54	1	11	50	4	13.8	36	8	26.7	54	1	15	54
Search time	1	17.1	90	0	5	38	0	8	30	1	18	90	0	9.8	90
Egress time	1	10.1	68	1	5.2	23	1	8	30	1	7	23	1	8.7	68
Charged on-street															
Access time	1	22.3	50	10	16.7	35	5	8.3	15	10	28.6	50	1	17.1	45
Search time	0	13.1	90	1	7.4	23	0	11.8	30	0	13.6	90	0	11.6	68
Egress time	1	9.9	93	1	5.9	15	1	7.5	15	1	6.9	30	1	9.5	93
Fee	0	1	8	0.1	0.5	2.7	0.2	0.7	2	0.3	1.6	8	0	0.7	5.3
Charged off-street															
Access time	3	24.2	54	0	10.6	36	2	14.9	48	3	23.8	54	0	17.1	54
Search time	0	8.1	60	0	2.9	20	0	7.2	45	0	7.2	20	0	6.2	60
Egress time	0	9.6	62	0	4.9	36	1	6.1	25	1	8.6	23	0	7.2	62
Fee	0	1.1	5.3	0	0.6	6.7	0	0.8	4	0.3	1.4	4	0	0.8	6.7
Multistory															
Access time	1	23.4	50	0	12.2	35	3	15	30	1	24.8	50	0	16.9	50
Search time	0	7	60	0	5	20	0	7.7	30	0	5.3	30	0	6.9	60
Egress time	1	6.9	30	0	4.2	30	1	4.6	15	1	6.6	30	0	5.4	30
Fee	0.2	1.9	10	0	0.7	5	0.2	0.9	2.5	0.2	2.6	10	0	1	5
Illegal															
Access time	1	22.9	54	0	10.9	50	2	14.4	48	1	24.9	54	0	16.2	54
Search time	0	3	5	3	3	3	1	2.9	3	0	2.9	5	1	2.9	3
Egress time	1	3.2	5	2	3.5	5	1	3.2	5	1	3.3	5	1	3.3	5
Expected fine	0	9.9	100	0	9.4	50	0	8	36	0	9.3	50	0	9.4	100

Table 4.2 Description of stated choice data sets

			Location			Activity	
		Overall	Birmingham	Sutton Coldfield	Coventry	Work (FT & PT)	Shopping and errands
	Number of respondents	298	137	89	72	51	247
Number of times available	Number of observations	1335	675	366	294	233	1102
Number of times available	Free on-street	498	254	130	114	124	374
Number of times available	Charged on-street	283	199	48	36	59	224
Number of times available	Charged off-street	964	448	286	230	128	836
	Multistory	925	449	268	208	155	770
	Illegal	1335	675	366	294	233	1102

4.4 METHODOLOGY

4.4.1 Data Rearrangement

Given the differences in scale and interpretation, it seems inappropriate to treat the cost for legal parking and the expected fine for illegal parking in the same way. Also, because respondents were presented with the overall fine level along with the probability of being caught (rather than the expected fine), the differences across respondents in their evaluation of this information can be guaranteed to be more important than is the case with the fixed-parking-fee attribute, given the differences in evaluations of risk. It was thus decided to treat the two attributes separately by using a cost parameter for legal forms of parking and a penalty parameter for illegal parking. This is not only more consistent with the real-world meaning of the two attributes, but is also helpful in the interpretation of the estimated values for the coefficients associated with the two parameters.

4.4.2 Choice of Model

The use of discrete choice models in transportation research has increased rapidly over the past three decades. These models are designed for the analysis of the choice between mutually exclusive alternatives. The choice probability of an alternative is a function of the relative utility of that alternative (compared to that of all other available alternatives), calculated as a function of the attributes of the alternatives and the tastes of the decision maker. For various reasons, including modeling uncertainty and the presence of nonmeasurable attributes, only part of an alternative's utility is observed, and the distributional assumptions regarding the unobserved part determine the structure and behavior of the resulting model (cf. Train, 2003).

Originally, most applications were based on the use of the multinomial logit (MNL) model (cf. McFadden, 1974). The MNL model has important advantages in terms of ease of estimation offset by certain shortcomings, notably in the form of inflexible substitution patterns when faced with correlated error terms in the data. Several alternative model forms have been proposed to address these problems, with the most prominent choice being the nested logit (NL) model (Daly and Zachary, 1978; McFadden, 1978; Williams, 1977), which improves flexibility by nesting similar alternatives together. Recently, the use of an even more flexible model form, the MMNL model, has increased dramatically, mainly thanks to improvements in the efficiency of simulation-based estimation processes which are required when using this model form. The crucial advantage of this model over other logit type models is that it allows for random taste variation across decision makers on top of any deterministic taste heterogeneity. This enables the MMNL model to give a more accurate representation of real-world behavior than its fixed-coefficients counterparts, which are limited to explaining taste heterogeneity in a deterministic way. Furthermore, the MMNL structure allows researchers to account explicitly for the serial correlations arising between repeated choice observations in the case of panel data.

The number of applications using the MMNL model has increased steadily the past few years. For some recent examples, see Algers et al. (1998), Train (1998), Revelt and Train (1998, 2000), Brownstone and Train (1999), and Hess et al. (2006). For a more detailed discussion of the power and flexibility of the MMNL model, and comparisons with other model forms, see McFadden and Train (2000) or Munizaga and Alvarez-Daziano (2001). For a discussion of the use of SC data in MMNL models, see Brownstone et al. (2000).

To the authors' knowledge, the MMNL structure has not been exploited in the modeling of parking-type choice. There have, however, been a number of studies using basic, nonsimulation-based discrete choice models in this area of research. For example, logit type models have been used by Ergün (1971) in the modeling of the choice of parking location and by Spiess (1996) in the modeling of parking lot choice in a park-and-ride context. Another application using the MNL model for the modeling of parking location is given by Teknomo and Hokao (1997), whereas Hunt (1988) uses NL models in the modeling of parking type as well as location. Finally, Bradley et al. (1993) use an NL model to predict changes in mode and parking-type choice resulting from changes in parking policies in major cities. Other applications have focused on the effects of parking availability on more general travel behavior.

For example, Hess (2001) uses an MNL model to assess the impact of the availability of free parking on mode choice and parking demand for work-related travel, whereas Hensher and King (2001) use an NL model to analyze the effects of parking cost and availability (by location) on the choice between car and public transport for journeys to the central business district.

The above discussion has shown that there have been a number of applications using basic discrete choice models in the analysis of parking behavior; however, there has been a distinct lack of applications using more advanced model forms, such as the MMNL model. Although the attributes of parking options have been used as explanatory variables in MMNL models (for example Bhat and Castelar, 2002), an important avenue of research remains unexplored in the actual MMNL modeling of parking behavior, with parking-type choice being but one example. Indeed, the extra flexibility of allowing for random taste variation can potentially offer great benefits in this area of research, given, for example, the differences among travelers in their sensitivity to search time or egress time. An even more likely source for taste variation is the attitude of decision makers toward illegal parking and their appraisal of the risks involved.

4.4.3 Model Specification and Estimation

The MMNL model uses integration of the MNL probabilities over the (assumed) distribution of the random parameters included in the model (Train, 2003). Formally, the probability of decision maker n choosing alternative i is given by:

$$P(n,i) = \int L_i(\beta, z_n) f(\beta \mid \theta) d\beta \tag{4.1}$$

where z_n is a matrix of the attributes of the alternatives as faced by decision maker n, and where the function $L_i(\beta, z_n)$ represents the conditional (on β) MNL choice probability given by:

$$L_i(\beta, z_n) = \frac{e^{\beta' z_{ni}}}{\sum_{j=1}^{I} e^{\beta' z_{nj}}} \tag{4.2}$$

where I gives the total number of alternatives in the choice set, and where z_{ni} is the vector of attributes of alternative i as faced by decision maker n. The vector β varies decision makers and reflects the idiosyncratic aspects of decision maker n's preferences; these terms are distributed in the population with density $f(\beta|\theta)$, where θ is a vector of parameters to be estimated that comprises, for example, the population mean and standard deviation of the single coefficients contained in vector β. In general, two parameters are associated with each randomly distributed coefficient, representing the mean and spread in the coefficient's values across the population.

The aim is to find optimal values of θ for the population used in the sample. For this, the likelihood function of the observed choices is maximized with respect to θ. Formally, with $i_{(n)}$ giving the alternative chosen by decision maker n, the likelihood function with N decision makers is given by:

$$L_N = \prod_{n=1}^{N} p(n, i_{(n)}) = \prod_{n=1}^{N} \left(\int L_{i(n)}(\beta, z_n) f(\beta \mid \theta) d\beta \right) \tag{4.3}$$

In the case of SC data (where we have multiple hypothetical choice situations per individual), the above formula needs to be adapted. Notably, for each respondent, the integral of the conditional choice probability $L_{i(n)}(\beta, z_n)$ over the distribution of β is replaced by an integral of the conditional choice probability of the sequence of observed choices for this individual, where the conditional choice probability of this sequence is given by the product of the conditional choice probabilities of the individual choices. Formally, with $T_{(n)}$ giving the number of choices observed for respondent n, and $i_{n(t)}$ representing respondent n's choice in the tth choice situation (with a corresponding explanation for $z_{n(t)}$), we have:

$$L_N = \prod_{n=1}^{N} \left(\int \left(\prod_{t=1}^{T(n)} L_{i_{n(t)}}(\beta, z_{n(t)}) f(\beta \mid \theta) \right) d\beta \right) \tag{4.4}$$

The MMNL model is calibrated by maximizing Equation (4.4) (Equation (4.3) in the case of cross-sectional data) for θ, thus finding the optimal values for representing the behavior observed in the sample; for optimization reasons, working with the log-likelihood is generally preferable (cf. Train, 2003). The maximization of this log-likelihood function clearly requires the calculation of the individual choice probabilities, respectively the choice probabilities of the observed sequences of choices for the different respondents. However, in the case of the MMNL model, the integrals representing these choice probabilities do not, in general, have a closed-form solution and need to be approximated, by simulation for example. For this, the value of a given integrand is calculated for a high number of draws from the relevant random distributions, and the average of these values over the set of draws is used as an approximation, where the precision of this approximation increases with the number of draws used. For more details on the consistency and efficiency of this simulation approach, see Train (2003). Given that the use of a high number of pseudorandom draws in simulation is computationally expensive, researchers are turning their attention increasingly to using alternative approaches known as quasirandom sequences. These sequences provide more uniform coverage of the area of integration than pseudorandom sequences, leading to a lower requirement in the number of draws, reducing the computational cost of the estimation and application of the MMNL model. For details on these approaches, see Bhat (1999), Train (2003), and Hess et al. (2003, 2006). In the present analysis, an approach known as the Halton sequence was used (Halton, 1960). The use of these sequentially constructed sequences has been observed to lead to important gains in simulation and estimation performance (Bhat, 1999, Train, 1999) and has been used successfully in the field of transportation research (Bhat, 2000), as well as in many other areas of economics.

In the present analysis, a total of 10 coefficients could be used; these are the coefficients associated with the five attributes of the alternatives (access time, search time, egress time, parking fee, and expected fine for illegal parking) and five alternative-specific constants (ASC) for the five parking options. For reasons of identification, one of the ASCs needs to be normalized to a value of 0. To minimize any loss of information (which would increase the error term in the model) in an MMNL model, the ASC with the least amount of variability across decision makers should be selected for normalization (cf. Hensher and Greene, 2001). In the present context, this was (for all subsamples) found to be the ASC for the free on-street alternative; this ASC was thus set to 0, such that the estimated values for the four remaining ASCs capture the net impact of unmeasured variables (including general attitude) on the respective alternatives' utilities relative to the free on-street option. A total of nine coefficients were thus used in the model. Given that two parameters are associated with each random coefficient, a maximum of 18 parameters, therefore, would need to be estimated.

4.4.4 Choice of Random Distributions

When using the MMNL model, an important question arises as to what distributions should be used for the coefficients. While the commonly used Normal distribution is a valid choice for a large selection of coefficients, the absence of constraints on the sign of the random variates makes it an inappropriate choice in the presence of an *a priori* assumption about the sign of a coefficient (for example, negative cost coefficients). The most commonly used distribution for such coefficients is the Lognormal distribution; this leads to positive coefficients, such that the sign of any undesirable attributes needs to be reversed to guarantee that increases in attribute values lead to decreases in utility.

While the Lognormal distribution has been used successfully in some MMNL applications (Bhat, 1998; Hess et al., 2006), it can occasionally lead to poor convergence and problems with unreasonably large parameter values, especially for the measure of spread (cf. Train, 2003). A solution to the latter problem is to use distributions that are bounded on both sides. Such distributions include the uniform and triangular distributions (Hensher and Greene, 2001) as well as the more advanced S_B distribution (cf. Train and Sonnier, 2003). For a more detailed discussion of existing approaches and past experience, see Train (2003).

It seems crucial to point out the importance of this issue of choice of distribution due to the potential effects of erroneous distributional assumptions on modeling results and policy decisions. Although the issue has been discussed in detail by some authors (Train, 2003; Hensher and Greene, 2001; Hess et al., 2005), it is still ignored by many authors, putting them at risk of producing seriously misleading results. In the current analysis, the Normal distribution could safely be used for the ASCs associated with the types of parking. In fact, the Normal distribution is the perfect choice for these coefficients. It allows for positive and negative values, thus reflecting the different attitudes, notably to illegal parking, observed across decision makers (where these attitudes form part of the unobserved variables whose impact is captured by the ASCs). For the remaining five coefficients, which reflect the sensitivity to costs in terms of time and money, the use of the Normal distribution cannot, in general, be justified, whereas strictly negative values normally would be expected for these coefficients. A non-signed distribution can thus lead to misleading results and potentially incorrect policy implications, so that, in the presence of significant taste variation, ideally a bounded distribution would be used.

Good results were obtained with the use of lognormally distributed values for the coefficients associated with access time, search time, egress time, and parking fee. However, problems with significant overestimating of the standard deviation arose when using the Lognormal distribution for the coefficient associated with the expected fine for illegal parking (in one example, this distribution produced a mean of 5 and a standard deviation of 500). Except for the three smaller data sets (Sutton Coldfield, Coventry, and work trips), however, the use of a fixed coefficient resulted in seriously underestimated coefficients and poor model fit, signaling the existence of significant levels of taste variation across decision makers. Experiments using distributions bounded on either side led to various problems, including slow convergence and poor model fit. On the other hand, good model fit, along with realistic parameter values (low probability of wrongly signed coefficients), was obtained when using normally-distributed coefficients. Although this is not fully consistent with the recommendations made regarding the use of the Normal distribution for coefficients for which assumptions exist about the sign, it was decided to forego these recommendations in those cases in which the probability of a wrongly-signed coefficient is at an acceptably low level. Any problems resulting from this were deemed to be less important than the poor model fit resulting from the assumption of no taste variation or the problems of poor estimates when using alternative bounded distributions.

4.5 RESULTS

In this section, the results produced by the models estimated on the various data sets are presented. The results are summarized in Tables 4.3, 4.4, 4.5, giving for each data set the results from the best fitting MMNL model alongside the results produced by an MNL model. This allows us to quantify the advantages offered by the MMNL model and shows the effect that the assumption of fixed coefficients (MNL) has on the values of coefficients (when compared with the mean values of their randomly distributed counterparts).

At this point, it seems worthwhile to point out a convention that was used in the presentation of the results for lognormally-distributed coefficients. Indeed, for these coefficients, the estimated parameters are the mean c and standard deviation s of the log of the coefficient. For ease of interpretation, the values presented in the table are in fact the actual mean and standard deviation of the lognormally-distributed coefficients given by:

$$\mu = \exp\left(c + \frac{s^2}{2}\right) \tag{4.5}$$

and:

$$\sigma = \mu\sqrt{\exp(s^2) - 1} \tag{4.6}$$

As the t-test values generated in the estimation are for the original parameters of the distribution, they do not relate directly to the values reported in the tables. Both mean and standard deviation are functions of c and s, therefore, it is important that both reported t-values are statistically significant. This differs from the case of normally-distributed parameters, where the t-test value of the standard deviation is of higher importance in the search for random taste heterogeneity. Also, for ease of interpretation, the sign of the mean values of lognormally-distributed coefficients was reversed in the tables to reflect the negative impact of the associated attributes on the utility of an alternative.

Aside from the implied values of time, the tables also show the ratio of the parking fee coefficient against the expected fine coefficient. With one exception, this ratio is strictly greater than 1, suggesting that money paid in parking fees carries a higher disutility than money paid in expected fines for illegal parking. This in turn suggests that when faced with the uncertain prospect of a parking fine, drivers behave as risk-prone decision makers.

While the use of randomly distributed coefficients in an MMNL model gives some indication of the variation in the coefficients across decision makers, the actual distributions matter most when looking at the distribution of ratios of coefficients, such as the values of time (willingness to pay). Here, it is important to incorporate the full distribution of the individual coefficient rather than working solely with the mean values of the coefficients. In the present analysis, there were no significant levels of correlation between the individual coefficients used in the model, such that a rather straightforward approach could be used in the calculation of the distribution of the values of time (no requirement to use a Choleski factor transformation). The approach starts by producing a high number of draws (100,000) for the various coefficients, using the distributional assumptions resulting from the model fitting exercise. For a given value of time statistic (access time, search time, and egress time), the required ratio was then calculated for each of the 100,000 pairs of draws used. As expected, the resulting statistics were all observed to follow a roughly lognormal distribution. Special care was required at this stage as the long tail of this distribution can lead to a high estimate of the mean and standard deviation of the value of time measures when based on the full sample of 100,000 draws. Therefore, it was decided to remove the upper two percentiles of the produced measures and to calculate the mean and standard deviations on the resulting sample of 98,000 measures (cf. Hensher and Greene, 2001). This method was used for the three measures of value time used, and for the four data sets in which at least one of the relevant coefficients follows a random distribution.

4.5.1 Overall Data Set

The first part of the analysis (Table 4.3) consisted of fitting a model to a data set containing information on all 298 respondents (1,335 observations). In the MMNL model, a Normal distribution was used for all four identified ASCs; the highly-significant standard deviations for these coefficients show the extent of taste variation in these coefficients, at least partly reflecting the differences in terms of respondents' attitudes toward the types of parking. It should be noted that in a model using SC data, the ASCs capture a range of effects, including both substantive effects relating to actual preferences and effects relating to the design of the SC survey. This should always be kept in mind when trying to infer information about actual agent behavior based on the estimates for these coefficients. Even so, the large negative value for the ASC associated with illegal parking can be seen to reflect the general law-abiding nature of the majority of the population at least to some extent. The fact that there is a 7.9 percent probability of the coefficient being positive further illustrates the extent of taste variation for this coefficient.

In terms of sensitivity to time, there is significant taste variation only for search time and egress time, leading to a fixed coefficient for access time, and lognormally-distributed coefficients for search time and egress time. A lognormal distribution was also used for the cost coefficient and, given the reasons mentioned in the earlier discussion, a Normal distribution had to be used for the expected fine coefficient. Although this does imply a probability of ~0.7 percent of a positive coefficient, this risk is a necessary evil in this case, because poor results were obtained with all of the alternative distributions.

A comparison with the MNL model shows important differences in model fit; with eight additional parameters, the MMNL model has increased the log-likelihood

Table 4.3 Modeling results on overall data set

Coefficient		Dist.	MMNL model	MNL model
Time				
Access	mean	F	−0.1174 (3.9)	−0.0311 (2.1)
	std. dev.		–	–
Search	mean	LN	−0.2159 (7.9)	−0.0575 (5.2)
	std. dev.		0.1604 (7.2)	–
Egress	mean	LN	−0.1855 (8.7)	−0.0850 (5.2)
	std. dev.		0.1961 (5.1)	–
Parking fee	mean	LN	−3.3069 (6.2)	−0.6306 (4.1)
	std. dev.		2.6249 (12.9)	–
Expected fine	mean	N	−2.2931 (4.9)	−0.3965 (4.4)
	std. dev.		0.9405 (5.0)	–
ASCs				
Charged on-street	mean	N	−3.1645 (3.1)	−1.0006 (3.2)
	std. dev.		3.9765 (5.1)	–
Charged off-street	mean	N	0.5975 (1.3)	−0.0373 (0.2)
	std. dev.		2.3408 (6.6)	–
Multistory	mean	N	0.6633 (1.2)	−0.0425 (0.2)
	std. dev.		3.5897 (7.0)	–
Illegal	mean	N	−6.4912 (4.6)	−2.2544 (5.7)
	std. dev.		4.5956 (4.5)	–
Values of time £h), mean (std. dev.)				
Access			3.23 (2.11)	2.95
Search			5.58 (5.08)	5.47
Egress			4.61 (4.90)	8.09
Ratio of fee against expected fine (means)			1.44	1.59
LL at convergence			−636.49	−882.65
Parameters estimated			17	9

by 246.16, whereas the 99.9 percent χ_8^2 significance limit for such a change is only 26.12. The effects of not allowing for random taste variation become most obvious in the negative coefficients associated with the ASCs for the charged off-street and multistory car park options (where the mean values in the MMNL model were positive). Finally, whereas the ratio of fee against expected fine is comparable, there are significant differences between the models in terms of implied values of search time and

egress time with egress time being valued in fact, more negatively than search time in the MNL model. For the MMNL model, the implied mean values of time show that access time is the least negatively-valued factor, whereas search time is valued more negatively than egress time. Also, the ratio between the mean coefficients for parking fee and expected fine shows that, as expected, the coefficient for parking fee is higher (in absolute value) than that for expected fine. The MMNL results produced with the parameters estimated on the overall data set show the effects of random taste variation on the implied values of time with high standard deviations for all three measures (partly as a result of variation in cost sensitivity).

4.5.2 Grouping by Location

The next step consisted of fitting separate models for the three locations considered (Table 4.4). For all three data sets, the use of the MMNL model again leads to highly-significant increases in the log-likelihood when compared with the MNL model.

There are significant differences in the number of observations across locations, and the number of observations in Sutton Coldfield and Coventry are so low (366 and 294 respectively) that significant taste variation could be identified for only a handful of coefficients. Indeed, the three time coefficients as well as the parking-fee and expected-fine coefficients had to be assumed to take on fixed values in these two models. Furthermore, although sufficient taste variation was identified in the Sutton Coldfield data set to allow the use of a Normal distribution for all four ASCs, in the Coventry data set this was possible only for the ASCs for charged off-street and illegal parking. The fact that the Coventry model uses only two randomly distributed coefficients, therefore, makes the increase in the log-likelihood by 59.39 even more remarkable given the 99.9 percent χ_2^2 significance limit of just 13.82.

The data set for Birmingham contained a sufficiently high number of observations (675) to reveal significant taste variation for the coefficients associated with search time (lognormal), parking fee (lognormal), and expected fine (Normal) as well as the ASCs for the charged on-street, charged off-street, and illegal parking options (all normally distributed). For the same reasons given in the previous section, a Normal distribution again had to be assumed for the expected fine coefficient. With the estimated parameters, there is a probability of a positive coefficient of ~0.4 percent, which is again acceptable given the circumstances. Even though the associated standard deviation is significant only at the 89 percent level, the use of a randomly distributed coefficient was justified by the fact that it leads to important gains in model fit over the use of a fixed penalty coefficient.

The implied values of time in the various MMNL models reveal significant differences across locations. The Birmingham model repeats the findings from

Table 4.4 Modeling results on data set using grouping by location

Coefficient		Birmingham			Sutton Coldfield			Coventry		
		MMNL model		MNL model	MMNL model		MNL model	MMNL model		MNL model
Time		Dist.			Dist.			Dist.		
Access	mean	F	−0.1099 (3.6)	−0.0461 (2.8)	F	−0.2977 (2.3)	−0.0590 (0.9)	F	−0.0413 (0.8)	0.0483 (1.7)
	std. dev.		—	—		—	—		—	—
Search	mean	LN	−0.1621 (7.3)	−0.0500 (4.1)	F	−0.2842 (3.1)	−0.1148 (2.9)	F	−0.0514 (1.1)	−0.0700 (2.2)
	std. dev.		0.1206 (2.7)	—		—	—		—	—
Egress	mean	F	−0.1470 (4.6)	−0.0888 (4.8)	F	−0.2367 (2.2)	−0.1228 (1.7)	F	−0.2149 (3.4)	−0.0825 (1.7)
	std. dev.		—	—		—	—		—	—
Parking fee	mean	LN	−2.6405 (4.8)	−0.8348 (3.9)	F	−4.5197 (2.0)	−1.1065 (2.3)	F	−1.4677 (1.9)	−0.4181 (0.9)
	std. dev.		1.3663 (4.7)	—		—	—		—	—
Expected fine	mean	N	−2.0081 (1.8)	−0.6374 (4.0)	F	−0.5318 (1.7)	−0.2665 (1.6)	F	−0.5409 (2.8)	−0.3875 (2.2)
	std. dev.		0.7547 (1.6)	—		—	—		—	—

ASCs									
Charged on-street	mean	N	−2.3387 (1.7)	−0.3999 (1.1)	N	−2.2543 (0.9)	−1.5265 (1.9)	F	−2.5016 (1.4)	−1.5475 (1.7)
	std. dev.		4.3671 (2.3)	—		3.2032 (2.0)	—		—	—
Charged off-street	mean	N	1.4107 (2.4)	0.4436 (1.5)	N	−0.1194 (0.1)	−0.0673 (0.2)	N	−0.6110 (0.6)	−0.6868 (1.4)
	std. dev.		2.0128 (4.4)	—		2.4714 (2.9)	—		5.5897 (4.0)	—
Multistory	mean	F	2.4515 (3.5)	0.8587 (2.4)	N	−1.4049 (1.1)	−0.4494 (0.9)	F	−0.6854 (0.9)	−0.6696 (1.3)
	std. dev.		—	—		6.8436 (3.9)	—		—	—
Illegal	mean	N	−5.0345 (2.7)	−1.8627 (3.4)	N	−11.876 (3.0)	−2.9469 (4.1)	N	−4.2277 (2.7)	−1.9558 (2.6)
	std. dev.		4.4304 (2.7)	—		6.9021 (2.6)	—		2.6371 (2.5)	—
Values of time (£/h), mean (std. dev.)										
Access			3.05 (1.37)	3.31		3.95	3.2		1.69	6.93
Search			4.42 (3.37)	3.59		3.77	6.23		2.1	10.05
Egress			4.09 (1.83)	6.38		3.14	6.66		8.79	11.84
Ratio of fee against expected fine (means)			1.32	1.31		8.5	4.15		2.71	1.08
LL at convergence			−319.65	−414.24		−157.47	−219.30		−148.15	−207.54
Parameters estimated			15	9		13	9		11	9

the previous section showing that search time is valued the most negatively, with access time being valued the least negatively. Again, the MNL model seems to significantly overestimate the value of egress time, which is ranked higher than search time. For Sutton Coldfield, the results from the MMNL model surprisingly show that access time is valued more highly than search time, which is valued higher than egress time. The values of time produced by the MNL model seem high for search time and egress time, and the ranking is also the opposite of that produced by the MMNL model. The differences in valuations of time are even more significant in Coventry, where egress time is ranked as the most negative ahead of search time and access time. The fact that the estimated coefficient for egress time is more than four times as important as that for search time shows the respondents' inherent dislike for the foot journey between the parking space and the final destination. This can be explained in part by noting that at the time of the SC survey, foot journeys to Coventry city center often involved walks through unattractive neighborhoods and the use of a large number of subways, making walking an unpleasant activity. Although the MNL model does, in this case, manage to retrieve the correct ordering of the values of time, it seems to overestimate the value of search time. This means that the model underestimates the ratio between the egress time and search time coefficients, which is all the more surprising because both coefficients were similarly kept fixed in the MMNL model. The reasons for this problem can be traced to the fact that the MNL model ignores the presence of variation in the charged off-street and illegal parking ASCs; this leads to high residual error in the model which then leads to poor estimation of some of the remaining coefficients. This is especially important in the case of the ASC for charged off-street parking. Here, the parameters estimated by the MMNL model lead to a high probability of a positive value for the associated coefficient, whereas in the MNL model, the sign of the coefficients is assumed to remain constant (negative) across the population.

Another observation that can be made from Table 4.4 is that there are significant differences across locations in the ratio of the parking fee coefficient against the expected fine coefficient. Respondents in Birmingham treat the parking fee coefficient in a similar way to the expected fine coefficient; given the differences in scale of the two associated attributes, this should lead to low utility for illegal parking. The high standard deviation of the ASC for illegal parking, however, leads to a significant probability of a positive impact by unmeasured variables on the utility of illegal parking; such a positive value of the ASC for illegal parking would lessen the negative impact of the coefficient associated with the expected fine. Special care must be taken in the interpretation of the values of the ASCs given the likely impact of the SC survey design on these estimates. Even so, the possibility of a positive ASC for illegal parking in the Birmingham model can be explained in part by the fact that at the time of the SC survey, fine enforcement in Birmingham was poor (ineffective court administration), making the risk of actual prosecution low. Whereas respondents thus react negatively to expected high fines, their general attitude toward illegal parking is less negative given their knowledge about the lax enforcement. Finally, the high ratio of fee against expected fine in Sutton Coldfield shows a low sensitivity of people toward expected fines, which is partly explainable by high average wages in that area. It should also be noted that in the case of Sutton Coldfield and Coventry, the MNL estimates lead to an underestimation of the ratio between the parking fee and expected fine coefficients. This is a result of the underestimated parking fee coefficient, which led to overestimated values of time in these models.

4.5.3 Grouping by Activity

Grouping the data by activity was the final part of the analysis (Table 4.5). The result was two data sets; one for full-time and part-time work (51 respondents, 233 observations), and one for shopping and errand trips (247 respondents, 1,102 observations). Given the low number of observations in the data set for work trips, it came as no surprise that significant taste variation was observed for only three coefficients; those associated with search time, egress time, and parking fee, all of which were assigned a lognormal distribution. It is interesting to note that significant variation was observed only for the ASCs in the models used for the smaller data sets in the grouping by location, whereas such taste variation is observed only for time and cost coefficients in the model for work trips. This could signal homogeneity in the working population in terms of the general attitude toward parking options (reflected by the overall impact of unmeasured attributes), but significant heterogeneity with regard to values of time (for example, sensitivity to lateness). With only three randomly distributed coefficients, the MMNL model still manages to increase the log-likelihood by 26.55, which compares well to the 99.9 percent χ_3^2 limit of 16.27. The *t*-value of the *s* parameter associated with the lognormal distribution of the egress time coefficient shows significance only at the 91 percent level, which is still acceptable. The implied values of time show that workers are most sensitive to egress time; this can be explained by various factors, including the monotony of the walk from the parking space to the final destination and a return journey that most workers have taken five times per week. Whereas the implied values of time in the MNL model are identical in terms of ranking and broadly similar in terms of ratios, the value of egress time seems especially high in the MNL model.

In the model for shopping and errand trips, a Normal distribution was used for all four ASCs and for the expected fine coefficient, along with a fixed access time coefficient and lognormal distributions for the search time, egress time, and parking fee coefficients, showing significant taste variation for all but one coefficient. The use

Table 4.5 Modeling results on data set using grouping by activity

Coefficient		Full-time and part-time work			Shopping and errand trips		
		MMNL model		MNL model	MMNL model		MNL model
Time		Dist.			Dist.		
Access	mean	F	−0.1563 (3.5)	−0.0513 (1.1)	F	−0.1004 (2.7)	−0.0283 (1.7)
	std. dev.		—	—		—	—
Search	mean	LN	−0.1674 (6.1)	−0.0632 (2.9)	LN	−0.4809 (3.7)	−0.0589 (4.5)
	std. dev.		0.1062 (2.0)	—		1.5974 (8.3)	—
Egress	mean	LN	−0.2338 (3.7)	−0.0925 (2.2)	LN	−0.2173 (5.9)	−0.0924 (5.0)
	std. dev.		0.1656 (1.7)	—		0.2370 (3.6)	—
Parking fee	mean	LN	−3.8206 (4.8)	−0.9727 (2.0)	LN	−4.5945 (3.4)	−0.5701 (3.5)
	std. dev.		2.4664 (4.9)	—		6.4869 (13.2)	—
Expected fine	mean	F	−1.8351 (3.2)	−0.8515 (5.2)	N	−5.7450 (4.1)	−0.2916 (3.2)
	std. dev.		—	—		2.7353 (3.9)	—

ASCs							
Charged on-street	mean	F	−2.6823 (1.0)	−2.7628 (2.1)	N	−1.5882 (1.6)	−0.8126 (2.5)
	std. dev.		—	—		2.2736 (2.5)	—
Charged off-street	mean	F	2.7228 (1.8)	0.2830 (0.5)	N	0.6057 (1.1)	−0.0913 (0.4)
	std. dev.		—	—		1.9604 (4.1)	—
Multistory	mean	F	4.1859 (2.6)	1.0614 (1.4)	N	0.8140 (1.2)	−0.2140 (0.9)
	std. dev.		—	—		4.4146 (5.4)	—
Illegal	mean	F	−1.8723 (2.4)	−0.8833 (1.3)	N	−8.7620 (2.7)	−2.8972 (5.5)
	std. dev.		—	—		5.0492 (2.6)	—
Values of time (£/h), mean (std. dev.)							
Access			3.31 (1.82)	3.16		3.37 (3.35)	2.98
Search			3.39 (2.64)	3.90		10.82 (20.55)	6.20
Egress			4.67 (3.80)	5.71		6.55 (8.76)	9.73
Ratio of fee against expected fine (means)			2.08	1.14		0.80	1.96
LL at convergence			−96.84	−123.39		−528.59	−731.22
Parameters estimated			12	9		17	9

of the Normal distribution for the expected fine coefficient does lead to a slightly high probability of a positive value (~1.8 percent); this can be seen as a necessary evil, however, given the poor performance when using a bounded distribution. The effects of using a fixed penalty coefficient are illustrated in the MNL model, which shows a ratio of fee against expected fine of 1.96, whereas, in the MMNL model, this ratio is 0.8, showing for the first time a higher sensitivity to expected fines than to parking fees. This should lead to low levels of illegal parking and is consistent with the notion that many shoppers do not mind using expensive parking as long as it is conveniently located. In this case, the MNL model significantly underestimates the negative impact of the expected fine level on the utility of illegal parking.

The MMNL model also shows that shoppers are more sensitive to search time than to egress time and access time. This can be explained in part by the low availability of parking spots during main shopping rush hours, resulting in a stressful search process that has a negative impact on the perceived overall quality of the shopping trip. As with all other data sets, the MNL model again assigns the highest valuation of time to egress time followed by search time and access time. Finally, with this data set, the MMNL model manages to increase the likelihood by 202.63 with eight additional parameters and with a corresponding 99.9 percent χ^2_8 significance limit of only 26.12.

4.6 SUMMARY

Our analysis has revealed the presence of significant taste variation in respondents' evaluation of parking options in terms of differences in valuation of the components of travel time, as well as in the impact of unmeasured variables and underlying preferences and the willingness to take risks when contemplating illegally parking. The extent of this random taste variation in the population was such that, for all data sets considered, the use of the mixed logit model resulted in important gains in model fit over the use of a simple multinomial logit model. This suggests that the use of the MMNL model can lead to important gains in accuracy in the modeling of parking behavior.

We have also highlighted some important issues in the specification of the MMNL model, especially with regard to the choice of distribution in the case in which an *a priori* assumption exists regarding the sign of some or all of the coefficients. Our analysis has reinforced earlier results with regard to occasional overestimation of the standard deviation of the coefficients when using a lognormal distribution.

The MMNL analysis has shown important differences in the valuation of the components of travel time across various locations and across different journey purposes. Indeed, while access time was valued the lowest in Birmingham and Coventry, it was valued higher than search time and egress time in Sutton Coldfield. Also, while egress time is valued second highest in Birmingham and lowest in Sutton Coldfield, it is valued higher than access time and search time in Coventry, to such an extent that the estimated value of egress time in Coventry is actually more than four times greater than the value of search time and more than five times greater than the value of access time. As for the differences across journey types, it was observed that workers value egress time the highest, whereas shoppers place more importance on search time, rating it over three times higher than workers do (when taking into account the full distribution of the value of time). There were significant differences between the MNL and MMNL model, both in terms of overall valuation levels as well as the ordering of different time components. To some extent, these differences can be seen as a reflection of potential bias in MNL results from not allowing for random taste heterogeneity.

Except for Sutton Coldfield, the results produced in the present analysis are broadly consistent with previous results, rating walk egress time higher than access time although the ratio between the two is generally below the lower limit of 2 suggested by Axhausen and Polak (1991) for this ratio. With the exception of Sutton Coldfield, our analysis confirms earlier findings rating search time higher than access time although the resulting ratio occasionally falls outside the interval suggested by Axhausen and Polak (1991), which ranges from 1.2 to 2.

As mentioned, it should be stressed that the data used in this research are rather dated, therefore, the results are not necessarily reflective of current parking behavior. Indeed, it can be assumed that the values of time would have been higher had a more up-to-date data set been used. Similarly, agents' attitudes towards illegal parking may have changed over time; the same observation could be made with regard to the perception of the relative levels of safety of the parking types. An important approach for further research is to use a flexible MMNL framework with a more up-to-date data set. It has also been suggested that other qualitative factors could be included in the evaluation of (especially off-street) parking options, such as the perceived risk of being mugged or hassled on the walk segment to and from the parking area. Also, as mentioned in 4.4.3, an important avenue for further research with this data set is a more explicit analysis of the correlation between SC replications.

However, as we have emphasized, the value of this work in the current context is not in the additional insight it provides into the processes of parking type choice, but the insight into the prevalence of heterogeneity in tastes. Although the latter may to some degree reflect heterogeneity in the incidence of SC-specific errors, rather than in underlying tastes *per se*, the overall magnitude of observed taste heterogeneity suggests that these effects do indeed play a significant role in parking choice and that analysts should account for them through suitable model specifications, such as MMNL. Finally, it should be noted that some of the heterogeneity in tastes identified by the MMNL models could be explained as a function of sociodemographic attributes such as income. Nevertheless, ongoing research has shown that the segmentation used in the present analysis (purpose and region) accounts for the majority of such deterministic taste variation, and that a significant remaining portion of heterogeneity can be explained only in a nondeterministic manner.

REFERENCES

Algers, S., P. Bergstroem, M. Dahlberg, and J. L. Dillen. 1998. Mixed logit estimation of the value of travel time, Working Paper, Department of Economics, Uppsala University, Sweden.

Axhausen, K. W. and J. W. Polak. 1991. Choice of parking: Stated preference approach. *Transportation*, 18: 59–81.

Bhat, C. R. 1998. Accommodating variations in responsiveness to level-of-service measures in travel mode choice modeling. *Transportation Research*, 32A(7): 495–507.

——— 1999. Quasi-random maximum simulated likelihood estimation of the mixed multinomial logit model. *Transportation Research*, 35B(7): 677–693.

——— 2000. A multi-level cross-classified model for discrete response variables. *Transportation Research*, 34B(7): 567–582.

Bhat, C. R. and S. Castelar. 2002. A unified mixed logit framework for modeling revealed and stated preferences: formulation and application to congestion pricing analysis in the San Francisco Bay Area. *Transportation Research*, 36B (7): 593–616.

Bradley, M. A., E. Kroes, and E. Hinloopen. 1993. A joint model of model/parking type choice with supply-constrained application, Proceedings of the 21st Annual Summer PTRC Meeting on European Transport, Highways and Planning, 61–73.

Brownstone, D. and K. Train. 1999. Forecasting new product penetration with flexible substitution patterns. *Journal of Econometrics*, 89: 109–129.

Brownstone, D., D. S. Bunch, and K. Train. 2000. Joint mixed-logit models of stated and revealed preferences for alternative-fuel vehicles. *Transportation Research*, 34B: 315–338.

Daly, A. and S. Zachary. 1978. Improved multiple choice models, *in* D. Hensher and M. Dalvi, eds., *Identifying and Measuring the Determinants of Mode Choice*. London: Teakfields.

Ergün, G. 1971. Development of a downtown parking model. *Highway Research Record*, 369: 118–134.

Halton, J. 1960. On the efficiency of certain quasi-random sequences of points in evaluating multi-dimensional integrals. *Numerische Mathematik*, 2: 84–90.

Hensher, D. A. and J. King. 2001. Parking demand and responsiveness to supply, pricing and location in the Sydney central business district. *Transportation Research*, 35A(3): 177–196.

Hess, D. B. (2001). The effect of free parking on commuter mode choice: Evidence from travel diary data. *Transportation Research Record*, 1753: 35–42.

Hess, S., J. W. Polak, and A. Daly. 2003. "On the performance of the Shuffled Halton sequences in the estimation of discrete choice models." Paper presented at the 30th European Transport Conference, Strasbourg.

Hess, S., M. Bierlaire, and J. W. Polak. 2005. Estimation of value of travel-time savings using Mixed Logit models. *Transportation Research Part A*, 39(2-3): 221–236.

Hess, S., K. Train, and J. W. Polak. 2006. On the use of a modified Latin hypercube sampling approach (MLHS) in the estimation of a Mixed Logit model for vehicle choice. *Transportation Research B*, 40(2): 147–163.

Hensher, D. and W. H. Greene. 2001. The Mixed Logit Model: The State of Practice and Warnings for the Unwary, Institute of Transport Studies, The University of Sydney, Sidney, Australia.

Hunt, J. D. 1988. "Parking location choice: Insights and representations based on observed behavior and hierarchical logit modeling formulation." Paper presented at the 58th Annual Meeting of the Institute of Transportation Engineers, Vancouver.

McFadden, D. 1974. Conditional logit analysis of qualitative choice behavior, *in* P. Zarembka, ed., *Frontiers in Econometrics*. New York: Academic Press, 105–142.

McFadden, D. 1978. Modeling the choice of residential location, *in* A. Karlqvist, L. Lundqvist, F. Snickars, and J. Weibull, eds., *Spatial Interaction Theory and Planning Model*. North-Holland, Amsterdam, 75–96.

McFadden, D. and K. Train. 2000. Mixed MNL Models for discrete response. *Journal of Applied Econometrics*, 15: 447–470.

Munizaga, M. A. and R. Alvarez-Daziano. 2001. Mixed Logit vs. Nested Logit and Probit models, Departamento de Ingeniería Civil, Universidad de Chile, Santiago, Chile.

Polak, J. W. and K. W. Axhausen. 1989. The Birmingham CLAMP Stated Preference Survey, second interim report to Birmingham City Council, Transport Studies Unit, Oxford University.

Polak, J. W., K. W. Axhausen, and T. Errington. 1990. The application of CLAMP to the analysis of parking policy in Birmingham City Centre. *Working Paper 554*, Transport Studies Group, Oxford University.

Polak, J. W. and P. Vythoulkas. 1993. An assessment of the state-of-the-art in modeling of parking behavior, *TSU Report 752*, Transport Studies Unit, University of Oxford.

Revelt, D. and K. Train. 1998. Mixed Logit with repeated choices: Households' choices of appliance efficiency level. *Review of Economics and Statistics*, 80(4): 647–657.

——— 2000. Customer-specific taste parameters and Mixed Logit: Households' choice of electricity supplier. *Working Paper No. E00-274*, Department of Economics, Berkeley: University of California.

Spiess, H. 1996. A logit parking choice model with explicit capacities. *Working Paper, EMME/2*. Support Center, Aegerten, Switzerland.

Teknomo, K. and K. Hokao. 1997. Parking behavior in central business district: A case study of Surabaya, Indonesia. *EASTS Journal*, 2.

Train, K. 1998. Recreation demand models with taste differences over people. *Land Economics*, 74: 185–194.

——— 1999. Halton sequences for mixed logit. Technical paper. Department of Economics, University of California, Berkeley.

———2003. *Discrete Choice Methods with Simulation*. Cambridge, MA: Cambridge University Press.

Train, K. and G. Sonnier. 2003. Mixed logit with bounded distributions of Partworths. Working paper, Department of Economics, University of California, Berkeley.

Van der Goot, D. 1982. A model to describe the choice of parking places. *Transportation Research*, 16A: 109–115.

Williams, H. 1977. On the formation of travel demand models and economic evaluation measures of user benefits. *Environment and Planning*, A9: 285–344.

5

MODELING DAILY TRAFFIC COUNTS: ANALYZING THE EFFECTS OF HOLIDAYS

Mario Cools
Transportation Research Institute
Hasselt University

Elke Moons
Transportation Research Institute
Hasselt University

Geert Wets
Transportation Research Institute
Hasselt University

5.1 INTRODUCTION

Two modeling philosophies for forecasting daily traffic counts are compared in this chapter. The premise is that successive traffic counts are correlated and, therefore, past values provide a solid base to forecast future traffic counts. In addition, it presupposes that daily traffic counts can be explained by other variables. Special attention is paid to the investigation of holiday effects. The analysis is performed on data originating from single inductive loop detectors collected in 2003, 2004, and 2005.

Results from both modeling philosophies show that weekly cycles predetermine the variability in daily traffic counts. The Box-Tiao modeling approach, which exploits the underlying proposition that explanatory variables can be used for forecasting future traffic counts, provides the required framework to quantify holiday effects. The results indicate that daily traffic counts are significantly reduced during holiday periods. When the forecasting performance of the modeling techniques was assessed, the Box-Tiao modeling approach out-performed the other modeling strategies, especially when a large forecast horizon was considered. Simultaneous modeling of travel motives and revealed traffic patterns is a key challenge for further research.

In our modern society, mobility is a driving force of human development. The motives for travel trips are not confined to work or educational purposes, but reach a spectrum of diverse goals. Mobility is more than a keystone for economic growth; it is a social need offering people the opportunity for self-fulfillment and relaxation (Ministerie van Verkeer en Waterstaat [Ministry of Transport, Public Works, and Water Management], 2004). The importance of mobility is recognized by governments at different policy levels. This is evidenced by the mobility plans formulated by government agencies such as the European Commission's white paper: "European Transport Policy for 2010: Time To Decide" (European Commission, 2004), and at Belgian regional level the "Mobility Plan Flanders" (Ministerie van de Vlaamse Gemeenschap [Ministry of the Flemish Community], 2001) and evidenced by the transportation research that is directly or indirectly funded by governments.

To lead an efficient policy, governments require reliable predictions of travel behavior, traffic performance, and traffic safety. A better understanding of the events that influence travel behavior and traffic performance will lead to better forecasts. Consequently, policy measures might be based upon more accurate data. Special holidays (for example, Christmas or New Year's Day), school holidays (July and August), sociodemographic changes, and weather can have an influence on mobility in several ways, as is illustrated by Figure 5.1 (Egeter and van de Riet, 1998). First, they can influence the travel market. This is the market in which the demand for activities and the supply of activity opportunities in space and time result in travel patterns. Second, these events can influence the transport market, the market where the demanded travel patterns and the supply of transport options come together in a transport pattern that assigns passenger and goods trips to vehicles and transport services. Finally, these events can have an effect on the traffic market, wherein the required transport patterns are confronted with the actual supply of infrastructure and their associated management systems, resulting in an actual use of the infrastructure revealed by the traffic patterns.

When the list of examples given in Figure 5.1 is considered, one notices that people might perform activities during holidays that vary from those performed

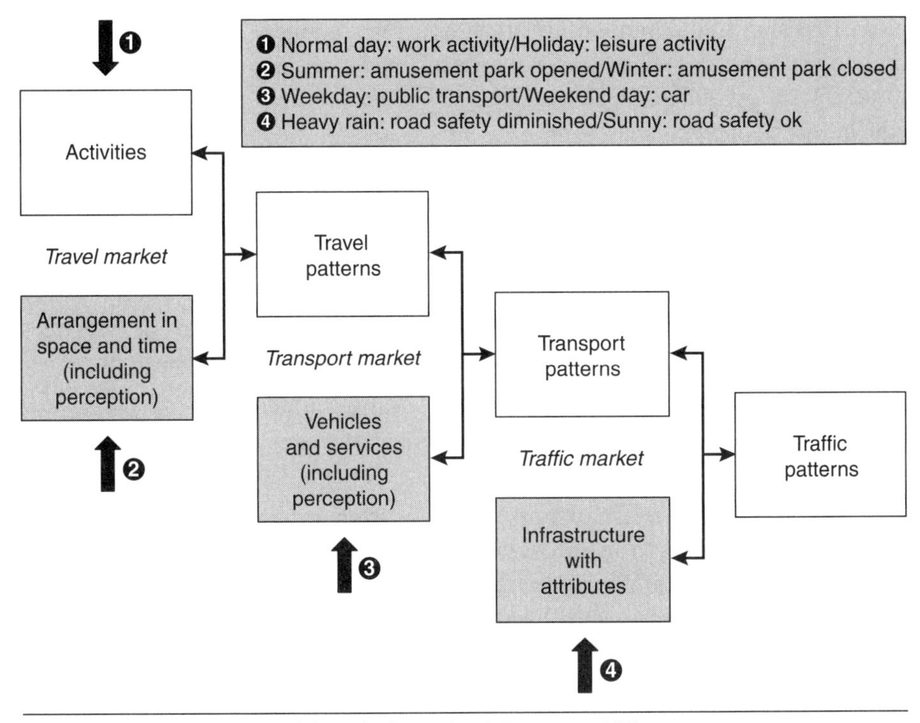

Figure 5.1 Three market models and effects that influence mobility

during normal days. During holidays, people can go to the beach, whereas during normal days, people go to work. Another effect indicated by Figure 5.1 is the closing of amusement parks during the winter. People wishing to visit the park during the winter obviously cannot and go ice-skating instead. These are but two examples of how holidays and seasonal effects influence the activities that people pursue and, in turn, how these activities have an impact on the travel market. Another example shows how mode choice can be influenced by the type of day and how this can have an impact on the transport market. The fourth illustration demonstrates how the environment can have an impact on the traffic market. Note that the list of examples in Figure 5.1 is not inclusive. It illustrates only three markets to illustrate that mobility can be influenced by various events.

This chapter pays special attention to the identification and quantification of the effects of holidays on daily traffic and to the prediction of future traffic volumes. A Box-Tiao model is used to quantify the holiday effects. A Box-Tiao model combines a regression model with auto-regressive (AR) and moving average (MA) errors, which creates the opportunity to build a model with desirable statistical properties and, thus, to minimize the risk of erroneous model interpretation (Van den Bossche et al., 2004). An overview of the data is presented, followed by a discussion of the imputation strategy that was applied. The methodology of the models used in the analysis is then explained, and the model outcomes and forecasts are presented. A general discussion and avenues for further research conclude the chapter.

5.2 DATA

The impact of holidays on daily traffic are analyzed by studying the effects on daily highway traffic counts. The dependent variable (daily traffic count) is explored, followed by a description of the covariates, referred to as *interventions* in Box-Tiao terminology.

5.2.1 Daily Traffic

The aggregated daily traffic counts originate from minute data of two single inductive loop detectors (one on every lane), located on the E19 Highway in the direction of Brussels in Vilvoorde (Belgium), collected in 2003, 2004, and 2005 by the Vlaams Verkeercentrum (Flemish Traffic Control Center). Figure 5.2 pin-points the traffic count location under study. The highway analyzed is one of the main access roads into Brussels and, thus, heavily travelled.

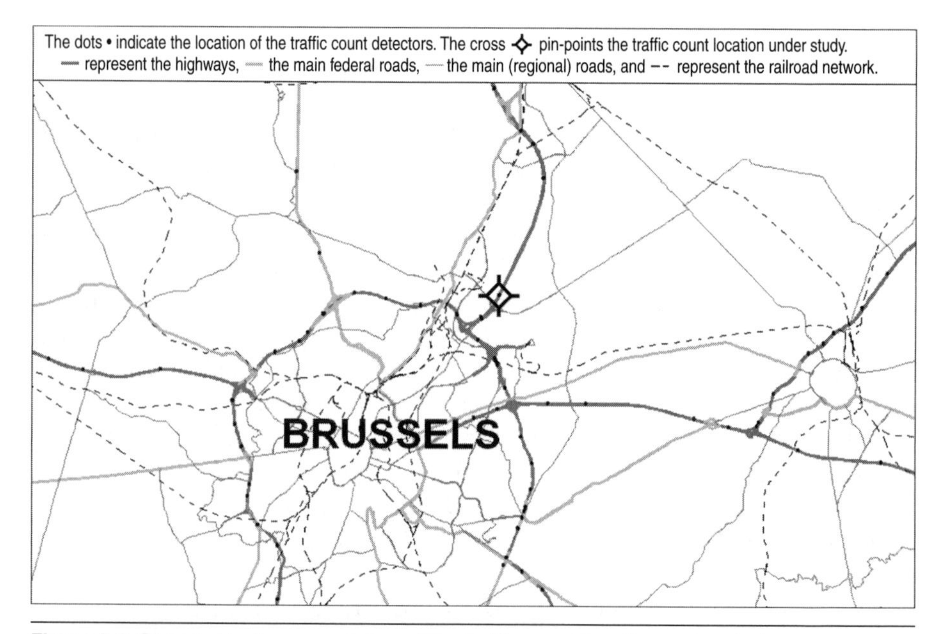

Figure 5.2 Geographical representation of the traffic count location under study

Every minute, the loop detectors output four variables: the number of cars driving by, the number of trucks, the occupancy of the detector, and the time-mean speed of all vehicles (Maerivoet, 2006). The number of cars and trucks are totaled for both detectors to yield a traffic count for each minute. The aggregation on a daily basis of these minute data can only be completed when there are no missing data that day. When some or all of the minute data are missing, a defendable imputation strategy must be applied.

About 30 percent of the days that were analyzed had no missing data, as shown in Table 5.1. Obviously, for these days, no imputation strategy needed to be applied. This, however, means that for the remaining 70 percent of the days, there were some (in 63.97 percent of the days) or many (in 7.75 percent of the days) of the minute-count data missing. When at least half the data—720 of the 1440 data points—were available, an imputation strategy was applied that is similar to the *reference days* method proposed by Bellemans (2003). When there were fewer than 720 data points available in a day, a more general imputation strategy was applied.

Bellemans (2003) assumed in his work the existence of an *a priori* known reference day that is representative of the day for which missing values have to be estimated. The imputed value is then calculated by scaling the reference measurements such that it corresponds to the traffic dynamics of the day under study. In his study, the scaling factor was the fraction of the measurement and the reference measurement in the previous minute.

The imputation strategy applied in this study utilizes the ideas of the reference days and the use of a scaling factor. The new measurements $x_{new}(t)$ are calculated in the following way:

$$x_{new}(t) = \delta x_{ref}(t) \tag{5.1}$$

where $x_{ref}(t)$ is the reference measurement at time t and δ the scaling factor. For determining the reference measurement, 21 reference days (seven days for each of

Table 5.1 Missing data analysis and corresponding imputation strategy

Quality Assessment	Number of days	% of all days	Imputation strategy
No min missing	310	28.28	no strategy
1-60 min missing	569	51.92	strategy 1
61-240 min missing	62	5.66	strategy 1
241-720 min missing	70	6.39	strategy 1
721-1439 min missing	34	3.10	strategy 2
Entire day missing	51	4.65	strategy 2
Totals	1096	100.00	

three holiday statuses) were used. For each reference day, the reference measurements were defined as the average of the modus, median, and mean of the available days that corresponded to the reference day. The average of these three measures of central tendency was taken, because each of them has its own unique attributes (central location, robustness, highest selection probability) and favoring one could obscure model interpretation. The scaling factor δ is calculated as follows:

$$\delta = \frac{\sum_{t=1}^{1440} d_t}{\sum_{t=1}^{1440} m_t}, \text{ where } d_t = \begin{cases} \frac{x(t)}{x_{ref}(t)} \Leftrightarrow x(t) \text{ not missing} \\ 0 \Leftrightarrow x(t) \text{ missing} \end{cases}, m_t = \begin{cases} 1 \Leftrightarrow x(t) \text{ not missing} \\ 0 \Leftrightarrow x(t) \text{ missing} \end{cases} \tag{5.2}$$

In the above equations, $x(t)$ is the measurement at minute t and $x_{ref}(t)$ is the reference measurement at minute t.

For this imputation strategy, a scaling factor was required to match the reference measurement to the day under study. When all, or nearly all, of the data points are missing, the scaling factor could not be calculated. In this case, the missing values are replaced by the reference measurements, which is equivalent to setting the scaling factor equal to one.

Circumspection is essential when applying imputation strategies, as imputation processes encompass the risk of distorting the distributions of the data and thus of biasing the results. The magnitude of the risk must be indicated and potential patterns of the missing data need to be analyzed. When the risk of distortion of the data is addressed, a thorough review of the minute data places the risk in the correct context. Of the 1,578,240 min (1,096 days multiplied by 1,440 min a day) that were aggregated on a daily basis, 157,272 min (9.97 percent) were missing. Communication errors (for example, due to system failures) account for the largest part (more than 90 percent) of the missing data problem. The remaining missing minute data were due to other reasons, including physical errors of the loop detectors, disturbances in the electronic systems of the substations, and inaccurate measurements.

When the imputation strategies are evaluated on the daily level, a first observation is that 80.20 percent (28.28 + 51.92) of the days contains at least 95.83 percent (more than 1,380 of the 1,440 data points) of the data points that day. Thus, the imputation strategy has nearly no effect on these days. For the days (4.65 + 3.10) that contained too little information (less than half of the 1,440 data points available), only a measure of central tendency was used as imputed value, taking into account the day type (which day of the week and whether a holiday). Because 12.05 percent (5.66 + 6.39) of the days between 50 percent and 95.83 percent of the data points were available, the scaling factor used for the imputation strategy was based upon a reliable amount of data.

It is important to stress that the imputation strategies applied use a measure of central tendency that takes into account the day of the week and the holiday status. Thus, the significance of these variables is not affected by the choice of the measure of central tendency. It is fair to recapitulate and infer that the implemented imputation strategies had no significant distorting effect on either the results or conclusions.

Figure 5.3 visualizes the aggregated daily traffic count data, taking into account the imputation strategies that were implemented. A similar pattern is visible over the three years. A drop in the number of passing vehicles at the beginning and end of each year is noticed and, during summer holidays, the intensity of daily traffic is clearly lower than during the other months.

5.2.2 Holiday and Day-of-Week Effects

A dummy variable was created to model the effect of holidays. Normal days were coded 0 and holidays were coded 1. The following holidays were considered: Christmas vacation, spring half-term, Easter vacation, Labor Day, Ascension Day, Whit Monday, vacation of the construction industry (three weeks, starting the second Monday of July), Our Blessed Lady Ascension, fall break (including All Saints' Day and All Souls' Day), and Remembrance Day. Note that for all of these holidays, the adjacent weekends were considered holidays also. For holidays

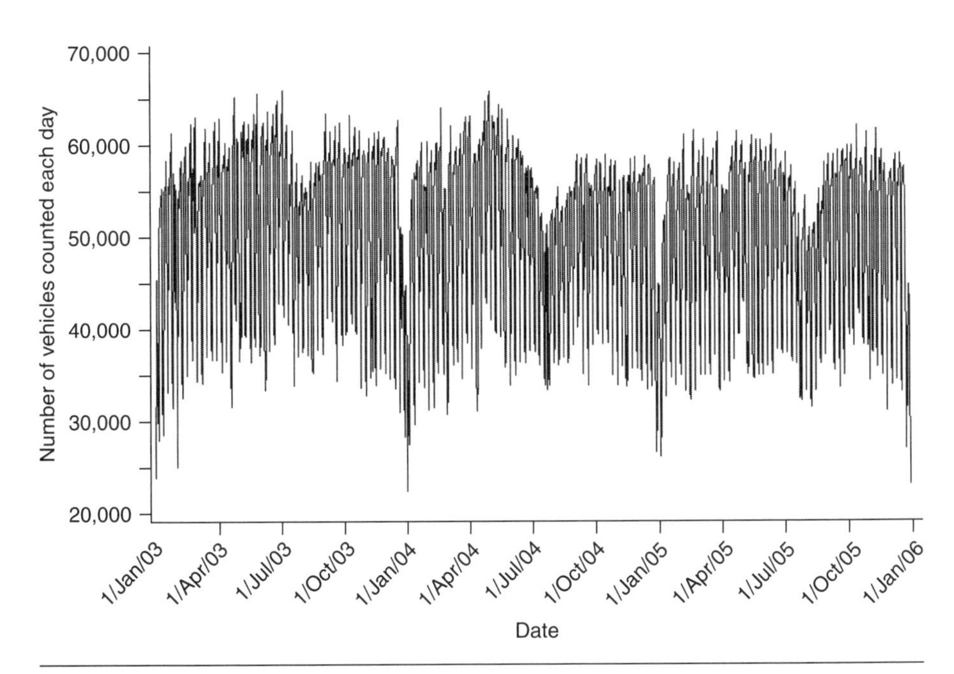

Figure 5.3 Evolution in time of daily traffic counts

occurring on a Tuesday or Thursday, the Monday and weekend before and the Friday and weekend after were also defined as holidays. Frequently people have a day off on those days and thus have a leave of several days which might be used for a long weekend or a short holiday. Note that for the imputation process, three holiday levels (normal days, holidays, and summer holidays) were considered. Notwithstanding, to obtain parsimonious estimation results, only two levels (normal days and holidays) were used in the estimation process.

Six dummy variables were created to model the day-of-week effect. Note that in general, $k - 1$ dummy variables must be created to analyze the effect of a categorical variable with k classes (Neter et al., 1996). Because there are seven days in a week, the first six days (Monday through Saturday) were each represented by one of the dummies, equal to 1 for the days they represent, and 0 elsewhere. The reference day was Sunday, therefore, all traffic counts that were collected on a Sunday, the corresponding six dummies were coded 0.

5.3 METHODOLOGY

This study explores two main philosophies for modeling daily traffic counts. The first is based on the fact that consecutive traffic counts are correlated and, therefore, that present and future values can be explained by past values. Two types of models that use this philosophy are investigated here: exponential smoothing and ARMA modeling. The second is the regression philosophy that postulates that the dependent variable—in this study daily traffic counts—could be explained by other variables. Various assumptions have to be met before the linear regression model yields interpretable parameter estimates, thus, the Box-Tiao model is also investigated. The latter is capable of taking into account dependencies between error terms. For an introduction to time series analysis, see Yaffee and McGee (2000); for a comprehensive overview of regression models, see Neter et al. (1996). Before elaborating on the modeling strategies that can be used for detecting patterns in traffic count data, spectral analysis is addressed.

5.3.1 Spectral Analysis

Spectral analysis is a statistical approach to detect regular cyclical patterns or periodicities. In spectral analysis, the data are transformed with a fine Fourier transformation and decomposed into waves of different frequencies (Tian and Fernandez, 1999; Fuller, 1976). The Fourier transform decomposition of the series is:

$$x_t = \frac{a_0}{2} + \sum_{k=1}^{m} [a_k \cos(w_k t) + b_k \sin(w_k t)] \tag{5.3}$$

where t is the time subscript, x_t are the data, n is the number of observations in the series, m is the number of frequencies in the Fourier decomposition ($m = \frac{n}{2}$ if n is even; $m = \frac{n-1}{2}$ if n is odd), a_0 is the mean term ($a_0 = 2\overline{x}$), are the cosine coefficients, b_k are the sine coefficients, and w_k are the Fourier frequencies ($w_k = \frac{2\pi k}{n}$).

Functions of the Fourier coefficients a_k and b_k can be plotted against frequency or wavelength to form periodograms, estimates of a theoretical quantity called a spectrum. The amplitude periodograms, also referred to as the periodogram ordinates, can then be smoothed to form spectral density estimates. The weight function used for the smoothing process, $W(\ )$, is often called the spectral window. The following simple triangular weighting scheme is used to produce a weighted moving-average estimate for the spectral density of the series: $\frac{1}{64\pi}, \frac{2}{64\pi}, \frac{3}{64\pi}, \frac{4}{64\pi}, \frac{3}{64\pi}, \frac{2}{64\pi}, \frac{1}{64\pi}$.

5.3.2 Exponential Smoothing

Simple exponential smoothing is a way of forecasting future observations by producing a time trend forecast in which the parameters are allowed to change gradually over time, and where recent observations are given more weight than observations in the past (Yaffee and McGee, 2000). The technique assumes that the data fluctuate around a reasonably stable mean. The formula for simple exponential smoothing is:

$$S_t = \alpha Y_t = (1-\alpha)S_{t-1} \tag{5.4}$$

where each new smoothed value S_t is computed as the weighted average of the current observation and the previous smoothed observation S_{t-1}. The magnitude of the smoothing constant α ranges between 0 and 1. If the constant equals 1, then the previous observations are ignored entirely. If the constant equals 0, then the current observation is ignored entirely, and the smoothed value consists entirely of the previous smoothed value; thus, as a consequence, all smoothed values are equal to the initial smoothed value S_0.

To accommodate the simple exponential smoothing model to account for regular seasonal fluctuations, the Holt-Winters method combines a time trend with multiplicative seasonal factors (SAS Institute Inc., 2004). The general formula for the multiplicative Holt-Winters model is:

$$\hat{Y}_{t+h} = (\mu_t + b_t h)S_{t-p+h} \tag{5.5}$$

where $\hat{Y}_{t+h}$ is the estimated response value for the time series at time $t + h$, h the number of periods into the forecast horizon, μ_t the permanent component at time t, b_t, the trend component at time t, S_{t-p+h}, the multiplicative seasonal component at time $t - p + h$, and p the periodicity of the seasonality (the number of periods in one cycle of seasons). Each of the three parameters (μ_t, b_t, S_t) is updated with its own exponential smoothing equation (Yaffee and McGee, 2000).

5.3.3 ARMA Modeling

Like exponential smoothing, the ARMA modeling approach also tries to explain current and future values of a variable as a weighted average of its own past values. In most cases, the model consists of a combination of an AR part and an MA part. When the series Y_t is modeled as an AR process AR(p), then Y_t can be expressed in terms of its own past values. Suppose Y_t is modeled as an AR process of order two, AR(2), then:

$$Y_t = c + \phi_1 B Y_t + \phi_2 B^2 Y_t + e_t \Leftrightarrow (1 - \phi_1 B - \phi_2 B^2) Y_t = c + e_t \tag{5.6}$$

where ϕ_1, ϕ_2 are the weights for the AR terms, c a constant, and e_t a new random term. In the above equation, B^i is used as a backshift operator on Y_t, defined as $B^i(Y_t) = Y_{t-i}$. When the series Y_t is modeled as an MA process MA(q), then Y_t can be expressed in terms of current and past errors, also called shocks. Suppose Y_t is modeled as an MA process of order two, MA(2), then:

$$Y_t = c + e_t - \theta_1 e_{t-1} - \theta_2 e_{t-2} \Leftrightarrow Y_t = c + (1 - \theta_1 B - \theta_2 B^2) e_t \tag{5.7}$$

where θ_1, θ_2 are the weights for the MA terms. In the cases that a series Y_t is modeled as a combination of an AR process of order p, AR(p), and an MA process of order q, MA(q), the combined process is called an ARMA(p,q) process. The model is then given by:

$$(1 - \phi_1 B - \phi_2 B^2 - \ldots - \phi_p B^p) T_t = c + (1 - \theta_1 B - \theta_2 B^2 - \ldots - \theta_q B^q) e_t \tag{5.8}$$

Note that the ARMA model is only valid when the series satisfies the requirement of weak stationarity. A time series is weakly stationary when the mean value function is constant and does not depend on time and when the variance around the mean remains constant over time (Shumway and Stoffer, 2000). If the variance of the series does not remain constant over time, a transformation, like taking the logarithm or the square root of the series, frequently proves to be a good remedial measure to achieve constancy (Neter et al., 1996). To achieve stationarity in terms of the mean, occasionally it is required to difference the original series. Successive changes in the series are then modeled instead of the original series. When differencing is applied, the ARMA model is called an AR*I*MA model—"*I*" indicates that the series is differenced.

5.3.4 Regression Modeling

Instead of modeling series Y_t as a combination of its past values, the regression approach tries to explain the series Y_t with other covariates. Formally, the multiple linear regression model can be represented by the following equation:

$$Y_t = \beta_0 + \beta_1 X_{1,t} + \beta_2 X_{2,t} + \ldots + \beta_k X_{k,t} + \varepsilon_t \tag{5.9}$$

where Y_t is the t-th observation of the dependent variable, and $X_{1,t}$, $X_{2,t}$, ..., $X_{k,t}$ are the corresponding observations of the explanatory variables. β_0, β_1, β_2, ..., β_k are the parameters of the regression model, which are fixed but unknown, and is the unknown random error (Neter et al., 1996). Estimates for the unknown parameters can be obtained by using classical estimation techniques. When the error terms are independently and identically normally distributed with mean 0 and variance σ^2, then the estimators for the parameters are called best linear unbiased estimators (BLUE).

5.3.5 Box-Tiao Modeling

When regression modeling is applied to time series, the assumption of independence of the error terms is often violated because of autocorrelation—the error terms are correlated among themselves. This violation of an underlying assumption of linear regression increases the risk for erroneous model interpretation because the true variance of the parameter estimates can be seriously underestimated (Neter et al., 1996). Box-Tiao modeling can be used to solve this problem of autocorrelation by describing the errors terms of the linear regression model by an ARMA(p,q) process. The Box-Tiao model can then be represented by the following equation:

$$Y_t = \beta_0 + \beta_1 X_{1,t} + \beta_2 X_{2,t} + \ldots + \beta_k X_{k,t} + \frac{(1-\theta_1 B - \theta_2 B^2 - \ldots - \theta_q B^q)}{(1-\phi_1 B - \phi_2 B^2 - \ldots - \phi_p B^p)} \varepsilon_t \quad (5.10)$$

where ε_t is assumed to be white noise. The parameters in this equation are determined using maximum likelihood estimation. Studies comparing the least-squares methods with the maximum likelihood methods for these kinds of models show that maximum likelihood estimation gives more accurate results (Brocklebank and Dickey, 2003). The likelihood function is maximized via nonlinear least squares using Marquardt's method (SAS Institute Inc., 2004). When differencing of the error terms is required to obtain stationarity, all dependent and independent variables should be differenced (Pankratz, 1991; Van den Bossche et al., 2004).

5.3.6 Model Evaluation

Because different types of models are considered to estimate the daily traffic counts, it is required that an objective criterion is used to determine which model performs better (Makridakis et al., 1998). The following criteria were used to determine the appropriateness of the models: the Akaike Information Criterion (AIC), the mean square error (MSE), and the mean absolute percentage error (MAPE). Note that the models were constructed on a training data set containing the first 75 percent of the observations. The remaining 25 percent of the observations make up the validation or test data set that can be used to assess the performance of the models by calculating the MSE and MAPE for the forecasts. The choice of these percentages is arbitrary, but it is common practice in validation studies (see Wets et al., 2000; Moons, 2005).

Traditionally, the AIC is defined as minus two times the log likelihood plus two times the number of free parameters. In the SAS procedure *Forecast,* which is used for the estimation of the exponential smoothing model, the AIC is approximated by the following:

$$AIC^* = n\ln\left(\frac{SSE}{n}\right) + 2k \qquad (5.11)$$

where n is the number of residuals, SSE is the sum of all squared errors and k is the number of parameters (SAS Institute Inc., 2004). Therefore, this approximation of the AIC is used to assess the model performance of the ARMA and Box-Tiao model. Models with a lower value for this criterion are considered more appropriate (Akaike, 1974). The root mean square error (RMSE) equals the square root of the sum of all squared errors (SSE) divided by its degrees of freedom, the latter of which are calculated by subtracting the number of parameters in the model from the number of observations. MAPE is defined as the average of the absolute values of the proportion of error at a given time.

5.4 RESULTS

The parameter estimates of the models are interpreted and the daily traffic counts are graphically displayed. A distinction is made between the predictions based on training data and those based on the test data. First, the results of the spectral analysis are highlighted.

5.4.1 Spectral Analysis

Figure 5.4 displays the plot of the spectral density estimates against the periods. This figure shows clearly that spectral density reached a local maximum in period 3.5 and global maximum in period 7. Thus, the spectral analysis has detected two regular cyclical patterns. The most prevalent, the global maximum of seven periods, can be interpreted as a weekly recurring pattern in the traffic count data. The periodicity of 3.5 periods is an indication of a half-week recurring cycle, yet it is much less dominant than the weekly cycle.

5.4.2 Holt-Winters Multiplicative Exponential Smoothing

The best Holt-Winters model, in terms of AIC^*, was obtained when a cycle of seven seasons (corresponding to the seven days of the week) combined with a linear trend was considered. In this model, nine (seven plus two) parameters

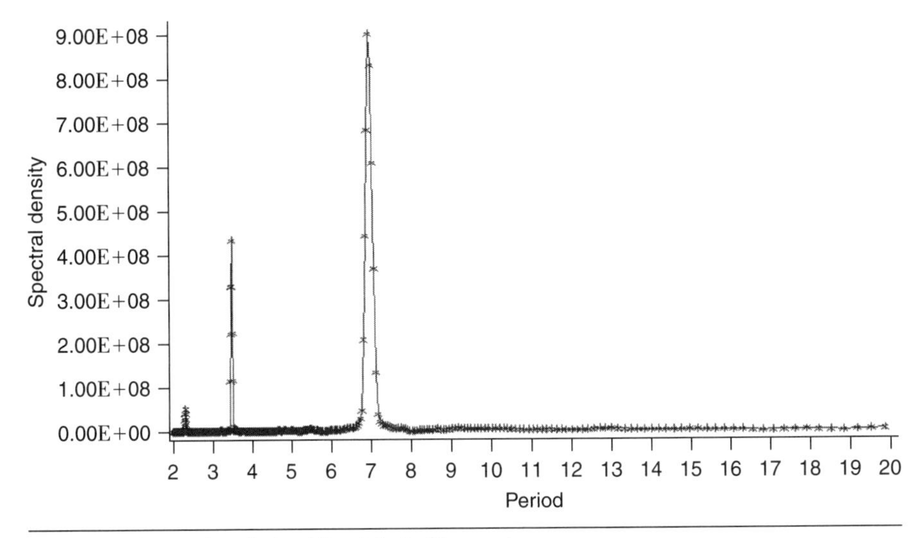

Figure 5.4 Spectral analysis of the daily traffic counts

had to be estimated: the parameter for the permanent component ($\hat{\mu}_1 = 47986$), the parameter for the linear component ($\hat{b}_1 = -192.3$), and the seven factors of the seasonal component. The estimated seasonal parameters are given as:

$$\begin{array}{ll} \hat{S}_1 = 1.125 & \hat{S}_5 = 0.691 \\ \hat{S}_2 = 1.156 & \hat{S}_6 = 1.035 \\ \hat{S}_3 = 1.159 & \hat{S}_7 = 1.092 \\ \hat{S}_4 = 0.742 & \end{array}$$

where $i = 1, 2, ..., 7$ represents the ordering of the seasonal parameters. The average of these seven parameters must be equal to 1 (Yaffee and McGee, 2000). Note that these seasonal factors correspond to the days of the week. Because the first observation in the data set was a Wednesday (January 1, 2003), the first seasonal factor also represents a Wednesday. Similarly, the other seasonal factors represent the other days of the week. Recall that the Holt-Winters method uses smoothing equations for updating parameters. The smoothing parameters for the permanent component and the linear component are given by $\alpha = \gamma = 0.106$, and the smoothing parameter for the seasonal component is given by $\delta = 0.25$. When the estimates for the seasonal parameters are compared, the difference between the components that correspond to the weekend days and the components that correspond to the weekdays is appealing. The results indicate that during weekend days the daily traffic count is much lower. This tendency can also be observed in Figure 5.5.

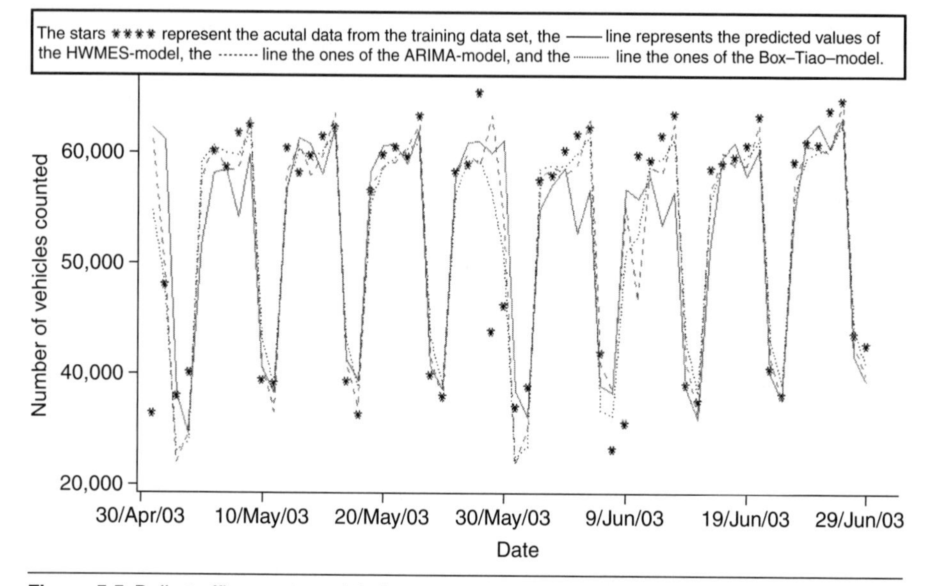

Figure 5.5 Daily traffic counts and their corresponding predicted values

5.4.3 ARMA Modeling

To obtain stationarity, the ARMA model was developed on differenced data. The autocorrelation function (ACF) and the partial autocorrelation function (PACF) of the residuals were investigated to determine which AR and MA factors were required to build the model. Let $\nabla_1 Y_t$ denote the first difference of the data ($Y_t - Y_{t-1}$) ; then the obtained model could be written as:

$$\nabla_1 Y_t = \frac{(1-0.926B)(1-B^4)(1-0.978B^7)}{(1-0.490B)(1-0.999B^4)(1-B^7)}\varepsilon_t \tag{5.12}$$

This model contains three multiplicative AR and three multiplicative MA factors. Notice that if the model is worked out, other AR and MA factors also play a role. When the parameter estimates for the ARMA factors are investigated, it can be seen that the estimates for the terms of the seventh order are close to or equal to 1. This is an indication of the weekly cyclic behavior also evidenced by the spectral analysis and the Holt-Winters model. The high parameters estimates for the ARMA factors of the fourth order might be evidence of some half-week recurring pattern in daily traffic counts. Recall that this half-week recurring pattern was also identified with the spectral analysis. The dependency on the previous day was much smaller, yet significant.

5.4.4 Box-Tiao Modeling

The classical linear regression modeling approach did not yield valid results because of the problem of autocorrelation of the error terms. As indicated in the methodological background, Box-Tiao modeling is an approach that can tackle this problem of autocorrelation. As in ARMA modeling, it was also required for Box-Tiao modeling to take the first difference of the data to obtain stationarity. Note that for both the ARMA model and the Box-Tiao model, the intercept was dropped from the equations. When differencing is done, the intercept is interpreted as a deterministic trend and that is not always realistic (Pankratz, 1991). The final error terms obtained were accepted to be *white noise*, according to the Ljung-Box Q* statistics (Ljung and Box, 1978). The final Box-Tiao model obtained is given by the following equation:

$$\nabla_1 Y_t = \begin{cases} -6042\nabla_1 X_{\text{Holiday},t} \\ +17073\nabla_1 X_{\text{Monday},t} + 18964\nabla_1 X_{\text{Tuesday},t} + 19426\nabla_1 X_{\text{Wednesday},t} \\ +19972\nabla_1 X_{\text{Thursday},t} + 21017\nabla_1 X_{\text{Friday},t} + 2460\nabla_1 X_{\text{Saturday},t} \\ +\dfrac{1-0.917B}{1-0.344B}\varepsilon_t \end{cases} \tag{5.13}$$

The six dummy variables to model the day-of-week effect and the dummy variable of the holiday effect are all significant (p-value < 0.0001) as seen in Table 5.2. This shows that the daily traffic counts are influenced by holidays. Interpretation of the parameter estimates is not straightforward because both the dependent and independent variables were differenced.

The parameter estimate for the holiday effect can be interpreted in the following way. When the holiday starts (when the differenced holiday dummy equals 1), the daily traffic count is 6042 vehicles lower than the day before. The day after the holiday (thus, when the differenced holiday dummy equals minus 1), the daily traffic count increases again with 6042 vehicles. Note that for all other days the differenced holiday dummy equals 0. For the interpretation of the parameter estimates for the day-of-week effects, the Wednesdays are taken as an example. On a Wednesday, the differenced dummy of the Wednesday-effect equals 1 and the differenced dummy of the Tuesday-effect equals minus 1. All other differenced day-of-week dummies equal 0 for a Wednesday. Thus, on a Wednesday, the traffic count is 462 (19426-18964) vehicles higher than the day before (obviously the Tuesday before).

5.4.5 Model Comparison

When the different models are compared, the weekly cyclic behavior was exposed by all three forecasting models. In the Holt-Winters Exponential Smoothing (HWES)

Table 5.2 Parameter estimates for the Box-Tiao model

Parameter	Estimate	Standard error	t-value	p-value
Moving average (Lag 1)	0.917	0.017	53.1	< 0.0001
Autoregressive (Lag 1)	0.344	0.040	8.6	< 0.0001
Holiday	−6042	451	−13.4	< 0.0001
Monday	17073	394	43.3	< 0.0001
Tuesday	18964	457	41.5	< 0.0001
Wednesday	19426	474	41.1	< 0.0001
Thursday	19972	471	42.4	< 0.0001
Friday	21017	453	46.4	< 0.0001
Saturday	2460	388	6.3	< 0.0001

model, this cyclicality was revealed by the seasonal component; in the ARMA model by the high estimates for the seventh order AR and MA factors; and in the Box-Tiao model by the clearly significant day-of-week effect. Differences between weekdays were also discovered by Weijermars and van Berkum (2005). In their work, they used cluster analysis techniques that revealed the differences.

To determine whether predicting daily traffic counts with other covariates—such as the holiday effect and the day-of-week effects—adds insight, criteria that assess the model fit are shown in Table 5.3.

When the different model comparison criteria are assessed, it is clear that the Box-Tiao model out-performs the other models, indicating that considering a holiday effect and day-of-week effects with a Box-Tiao model really adds insight into the cyclicality of daily traffic counts. Note that Liu and Sharma (2006) also stressed the importance of holidays on traffic phenomena. Figure 5.5 shows that the predictions

Table 5.3 Criteria for model comparisons

Criterion	Holt-Winters	ARMA	Box-Tiao
Comparison based on training data set (model-based criteria)			
AIC*	---	15,971	15,782
AIC	13,947	13,609	13,451
RMSE	4,809	3,961	3,591
MAPE	6.67%	5.48%	5.26%
Comparison based on test data set (forecast-based criteria)			
RMSE	32,784	4,862	4,184
MAPE	57.13%	6.78%	6.32%

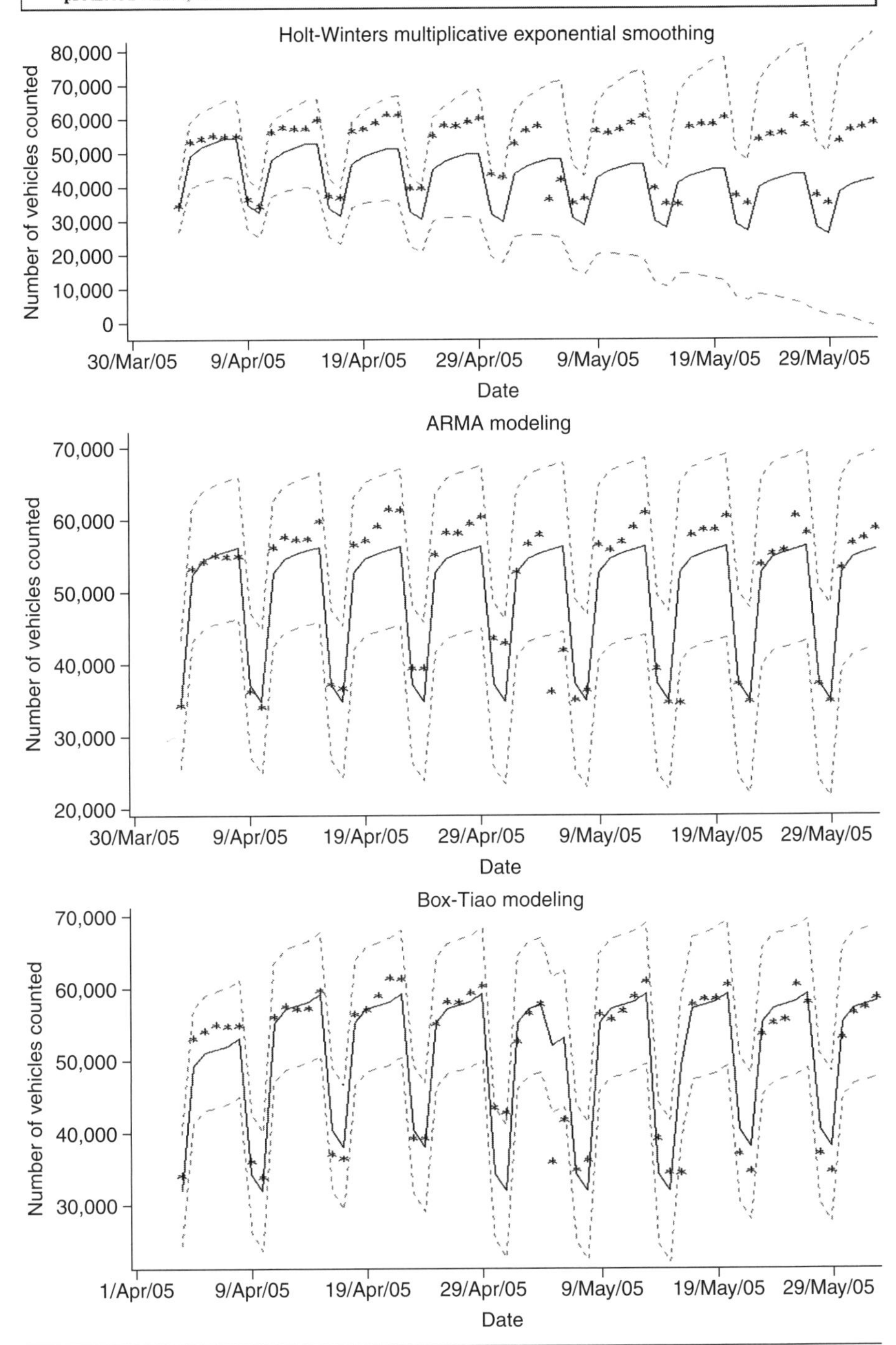

Figure 5.6 Daily traffic counts and their corresponding predicted values and confidence bounds

based upon the training data set are comparable for the three modeling strategies. Notwithstanding, all model-based criteria for these predictions indicate that the Box-Tiao models perform best. When the different models are validated on a test data set, the same conclusions can be formulated as with the training data, namely that the Box-Tiao model performs best. Figure 5.6 shows that the ARMA model also performs quite well, but the Holt-Winters model only performs well for a short forecast horizon. The RMSE and MAPE criteria demonstrate that the ARMA and Box-Tiao model approaches out-perform the Holt-Winters exponential smoothing model, favoring the Box-Tiao model.

5.5 SUMMARY

Different modeling approaches were considered to predict daily traffic counts. These techniques noted the significance of the day-of-week effects. Weekly cycles seem to determine the variation of daily traffic flows. On weekends, the daily traffic flows turn out to be lower than during the week. The Box-Tiao model approach demonstrated that during holidays, the daily traffic flows are significantly lower. When forecasting daily traffic flows is required, the Box-Tiao model appears to perform reasonably well. Smoothing techniques, such as the Holt-Winters, are to be avoided for predictions with a large forecast horizon. These findings can be used by policy makers to fine-tune current policy measures. More precise travel information can be provided, and the dynamic traffic management systems can be improved. In this way, the findings of the study contribute to achieving an important goal: more acceptable and reliable travel times.

The analysis of day-of-week and holiday effects in this study was done on the revealed traffic patterns. Generalization of the discussed results is possible when traffic patterns of other parts of the road network are analyzed. To get more insight into how holidays affect mobility, further analysis is required. The modeling techniques described here could be applied to data from national travel surveys to determine potential effects on travel behavior. Simultaneous modeling of both the underlying reasons for travel and revealed traffic patterns is a challenge for further research.

ACKNOWLEDGMENTS

The authors would like to thank the Vlaams Verkeerscentrum (Flemish Traffic Control Center) for providing the data used in this study.

REFERENCES

Akaike, H. 1974. A new look at the statistical model identification. IEEE transaction on automatic control, 19: 716–723.

Bellemans, T. 2003. "Traffic control on motorways." PhD Thesis. Katholieke Universiteit Leuven, Department of Electrical Engineering ESAT-SCD (SISTA).

Brocklebank, J. C. and D. A. Dickey. 2003. *SAS for Forecasting Time Series*, 2d ed. Cary, NC: SAS Institute.

Egeter, B. and O. A. W. T. Van de Riet. 1998. Systeemdiagram voor het beleidsveld vervoer en verkeer. Rapportnummer inro/VVG, 1998-02. TNO Inro/Rand Europe & TU Delf.

European Commission. 2004. White paper on European transport policy 2010: Time to decide. Luxemburg: Office for Official Publications of the European Communities.

Fuller, W. A. 1976. *Introduction to Statistical Time Series*. New York: John Wiley.

Liu, Z. and S. Sharma. 2006. Predicting directional design hourly volume from statutory holiday traffic. Transportation research record. *Journal of the Transportation Research Board*. 1968: 30–39.

Ljung, G. and G. Box. 1978. On a measure of lack of fit in time series models. *Biometrika*, 65(2): 297–303.

Maerivoet, S. 2006. "Modeling traffic on motorways: State-of-the-art, numerical data analysis, and dynamic traffic assignment." PhD thesis. Katholieke Universiteit Leuven, Department of Electrical Engineering ESAT-SCD (SISTA).

Makridakis, S., S. Wheelwright, and R. Hyndman. 1998. *Forecasting: Methods and Applications*, 3d ed. New York: John Wiley and Sons.

Ministerie van de Vlaamse Gemeenschap. 2001. Ontwerp mobiliteitsplan vlaanderen. Brussels: Departement Leefmilieu en Infrastructuur, Mobiliteitscel. (in Dutch).

Ministerie van Verkeer en Waterstaat. 2004. Nota Mobiliteit. Den Haag. (in Dutch).

Moons, E. 2005. "Modeling activity-diary data: Complexity or parsimony?" PhD thesis. Limburgs Universitair Centrum, Faculty of Science, Mathematics.

Neter, J., M. H. Kutner, C. J. Nachtsheim, and W. Wasserman. 1996. *Applied Linear Statistical Models*. Chicago: WCB/McGraw-Hill.

Pankratz, A. 1991. *Forecasting with Dynamic Regression Models*. New York: John Wiley & Sons.

SAS Institute Inc. 2004. *SAS/ETS 9.1 User's Guide*. Cary, NC: SAS Institute Inc.

Shumway, R. H. and D. S. Stoffer. 2000. *Time Series Analysis and Its Applications.* New York: Springer.

Tian, J. and G. Fernandez. 1999. Seasonal trend analysis of monthly water quality data. In Proceedings of the 7th Annual Western users of SAS software regional users group, 229–234.

Van den Bossche, F., G. Wets, and T. Brijs. 2004. A regression model with arma errors to investigate the frequency and severity of road traffic accidents. Proceedings of the 83th annual meeting of the Transportation Research Board. CDROM. Washington, DC: Transportation Research Board of the National Academies.

Wets, G., K. Vanhoof, T. Arentze, and H. Timmermans. 2000. Identifying deci sion structures underlying activity patterns: An exploration of data mining algorithms. *Transportation Research Record: Journal of the Transportation Research Board,* 1718: 1–9.

Wijermars, W. A. M. and E. C. Van Berkum. 2005. Analyzing highway flow patterns using cluster analysis. Proceedings of the 2005 IEEE Intelligent Transportation System Conference (ITSC). CDROM. Vienna, Austria.

Yaffee, R. A. and M. McGee, 2000. *Introduction to Time Series Analysis and Forecasting, with Applications of SAS and SPSS.* San Diego: Academic Press.

6

ISSUES WITH SMALL SAMPLES IN TRIP GENERATION ESTIMATION

Paul Metaxatos
Urban Transportation Center
University of Illinois–Chicago

6.1 INTRODUCTION

This chapter revisits issues facing transportation planners estimating and validating trip generation rates from small-scale household travel surveys. Three problems are addressed: (1) unusual observations, (2) small number of observations, and (3) no observations. Unusual observations are identified using traditional methods. Classification and regression tree (CART) analysis is proposed for the second problem. Finally, the third problem is addressed using row-column decomposition analysis. The methods are demonstrated using a small-scale household travel survey and are simple enough to be implemented with the resources available to transportation analysts, especially in smaller metropolitan planning organizations (MPOs).

Within the context of the federal transportation program, states and local agencies must have technical planning processes in place to program federal funds. As a result, transportation policy-making organizations called MPOs must develop and maintain transportation models to support various transportation- and land-use policies in every urbanized area with a population greater than

50,000. Such requirements raise concerns, particularly among smaller MPOs that lack the resources to undertake expensive primary data collection such as household travel surveys that are typically used in model development. Frequently, even if survey data are available, transportation professionals are requested to conduct complicated modeling tasks based on small samples. In such cases, the investigation of data-quality issues during the model development stage becomes critical.

Although several new approaches in travel modeling methods have emerged in recent years, the traditional four-step procedures (variants of the trip generation, trip distribution, mode split, and traffic assignment steps) are still in widespread use. Perhaps because the modeling community's focus has somewhat shifted to newer models, some previous questions about the reliability and stability of trip generation rates, particularly when sample sizes are small, remain unanswered. This chapter revisits the trip generation step of the conventional four-step modeling process and focuses on problems that are particularly troublesome when relatively small samples (on the order of only a few hundred observations) are used for data collection. We also focus on categorical trip generation models only, which are the ones used most often in practice, and for good statistical reason.

Difficulties encountered when using small surveys and methods proposed to alleviate them are illustrated using a small-scale household travel survey from Champaign–Urbana, Illinois. In the sections below, three specific issues are addressed: identification of outliers, reliability of small samples, and imputation issues in cases of missing information:

- *Identification of outliers:* When sample sizes are large, estimates are generally affected minimally by a suspect observation. However, with small sample sizes, a so-called *bad* observation (because of a mistake in recording data or a respondent giving misleading information) can play havoc with the quality of estimates and have a profound effect on forecasts. Such bad estimates need to be of concern only when they are influential. Influential observations that are also typically outliers are discussed in a later section.
- *Reliability of small samples:* Table 6.1 shows a typical (categorical) trip generation table. The total number of observations is 360 with the consequence that cells representing several categories of household sizes and numbers of workers have few observations. In such a case, reliability can be enhanced by combining table cells with like trip rates. A method for doing so is CART analysis. The procedure discussed later is well known in the statistical literature but has not been used widely in transportation.
- *Imputation:* Some cells might not receive an observation at all; or the analyst might find the observations too few or unreliable. In that case,

Table 6.1 Average number of trips per household (number of households in parentheses)

Number of workers	Household size			
	1	2	3	4+
0	6.28 (53)	9.22 (40)	16.00 (3)	8.00 (3)
1	6.09 (87)	10.42 (45)	10.94 (18)	9.06 (15)
2	–	9.5 (46)	10.81 (11)	15.46 (28)
3	–	–	11.00 (2)	11.83 (6)
4	–	–	–	10.66 (3)

–: indicates category not possible in this classification.

cell estimates might be imputed from observations in other cells. A procedure called *row-column decomposition analysis* is discussed later. Imputation can also be used in lieu of combining cells using CART, random forests, or other methods. However, such methods are not typically available to smaller MPOs.

Although the focus here is on obtaining estimates from small surveys, some of the discussion is germane to all trip generation efforts. Notice that the concern is not so much the overall size of the sample, but rather the size of the sample in each cell of the cross classification. Most applications of the four-step procedure are typically implemented using two-way classifications: number of adults in household and number of vehicles. Improvements can be made by increasing the number of variables in the classification under consideration. This has not been done because then the observations per cell would decline. The methods given in this chapter are used to address this issue.

6.2 STUDY AREA AND DATA

The Champaign–Urbana–Savoy urbanized area had a 2000 population of nearly 123,900 people (including University of Illinois students) who live in an area of more than 40 square miles (Champaign Urbana Urbanized Area Transportation Study [CUUATS], 2004). The area is located in east central Illinois approximately 135 miles south of Chicago.

The data used for the analysis in this chapter are based on the Champaign–Urbana–Savoy 2002 household travel survey conducted by the local MPO, CUUATS, in Illinois (Morocoima-Black and Kang, 2003). The survey was conducted to facilitate the development of a transportation model in support of the long-range transportation plan for the urbanized area.

A one-day home-interview survey collected data on the average weekday, local and regional personal travel made by residents of the Champaign–Urbana–Savoy urbanized area and was conducted from summer of 2002 to spring 2003. It had a 15 percent response rate and contained a sample of 362 usable households, approximately 0.7 percent of households in the study area. For comparison purposes, and given that sampling errors are dependent on absolute sizes of samples and not on percentages, the Chicago Area Transportation Study's 1990 Household Travel Survey obtained a sample of more than 19,000 usable households or approximately a 0.7 percent sample of households.

6.3 SMALL SAMPLES IN TRIP GENERATION

Small samples in trip generation occur for two reasons. The first is that simply the size of the overall survey sample is small. This situation may occur, for example, at smaller MPOs in which staff, budget, and other resources are too limited to obtain a sample size suitable to meet the survey objectives. As a result, population parameters (for example, mean trip rates per unit, modal proportions, and so on) that will be estimated based on the survey will be subject to greater sampling variability.

However, frequently the statistical analysis of survey data during trip generation modeling results in small samples. For example, in cross-classification analysis during model development, a table of family size by the number of workers in the household may have cells that have small samples of responding units (households, persons, and so on) or no samples at all. Moreover, higher dimension classifications would result in cells with even thinner samples.

To illustrate, consider the categorical (cross-classification) trip generation model in Table 6.1. This model estimates the average number of trips (trip rate) made by households using a two-way classification of the factors of family size and number of workers, each with several categories (for example, one-member households, two-member households or zero-worker households, one-worker households, and so on). Notice that several of the categories (cells) in Table 6.1 have a small number of responding households.

The problem is amplified when we expand a cross-classification to include additional categories or independent variables. For example, in a given travel survey, a four-by-five classification of family size and number of workers results in smaller samples than a three-by-four classification of the same variables. Similarly, adding more variables results in thinner samples. For example, a three-by-four-by-four classification of family size, number of workers, and vehicle availability has a smaller number of samples than a two-way classification of the first two variables.

For illustration purposes in Table 6.2, consider, in addition to the variables in Table 6.1, four levels of auto availability (*auto* is a generic term used to include cars, vans, and pick-up trucks) as a third classification variable (CUUATS, 2002). Clearly, there are several instances in this three-way classification with no households and several more with a small number of households. In Table 6.2 we have 56 possible cells (five worker categories times four household categories, times four vehicle categories, resulting in 80 cells with 24 cells not possible), which strains the ability to get reasonable numbers of observations within each cell.

6.4 IDENTIFICATION OF OUTLIERS

The first problem addressed is the identification of outliers that emerge when survey responses (observations) that involve trip-making activity stand out as

Table 6.2 Average number of trips per household (number of households in parentheses)

Workers	Vehicle availability	Household size			
		1	2	3	4+
0	0	5.85 (14)	ND (0)	ND (0)	ND (0)
0	1	6.60 (35)	10.14 (14)	8.00 (1)	9.00 (2)
0	2	5.00 (4)	8.66 (24)	20.00 (2)	6.00 (1)
0	3+	ND (0)	9.50 (2)	ND (0)	ND (0)
1	0	6.09 (22)	ND (0)	ND (0)	ND (0)
1	1	6.05 (59)	10.44 (18)	9.36 (11)	8.88 (9)
1	2	6.60 (5)	10.45 (24)	13.60 (5)	10.00 (5)
1	3+	6.00 (1)	10.00 (3)	13.00 (2)	6.00 (1)
2	0	–	8.00 (1)	ND (0)	ND (0)
2	1	–	9.00 (12)	12.66 (3)	15.00 (4)
2	2	–	9.92 (27)	9.25 (4)	15.85 (20)
2	3+	–	9.33 (6)	11.00 (4)	14.00 (4)
3	0	–	–	ND (0)	ND (0)
3	1	–	–	ND (0)	ND (0)
3	2	–	–	19.00 (1)	9.00 (2)
3	3+	–	–	3.00 (1)	13.25 (4)
4	0	–	–	–	ND (0)
4	1	–	–	–	ND (0)
4	2	–	– –	–	ND (0)
4	3+	–	–	–	10.66 (3)

–: indicates category not possible in this classification; ND: no data.

unusual. This problem is particularly vexing when the observation is the only one or one of very few in a cell (in a cross-classification table) and has an inordinate influence on trip rates. To counteract this, large numbers of observations would be required, especially in higher-dimension classifications. However, often this is not feasible in the smaller sample sizes typically employed in smaller towns.

To identify outliers in the CUUATS 2002 household travel survey, we employed traditional methods found in standard statistical textbooks (see Sen and Srivastava, 1990). If we consider the trip generation model in Table 6.1 as a regression problem (see Thakuriah et al., 1993 for a discussion about the equivalency between categorical and regression models) with trip rates as the dependent variable and the number of workers and family size as the independent variables, then it is reasonable to assume that outliers would exhibit numerically large residuals. However, not all influential points would have large residuals. Therefore, we focused our attention on households with a large residual or those located far away from other households in the space of the independent variables (number of workers and family size in this case).

There are numerous measures available in the literature for identifying outliers (Sen and Srivastava, 1990). Here, we examined the Studentized residuals (see Sen and Srivastava, 1990, for an exact definition), resulting from the regression above. It can be proved that this quantity has a *t* distribution when the errors are Gaussian and has a near *t* distribution under a wide range of circumstances. A Studentized residual is a standardized measure of the distance between a case (an observation) and the model estimated on the remaining cases. Therefore, it can be used as a test statistic to determine whether a case belongs to the model.

For a search of influential points, we examined the measure DFFITS (Sen and Srivastava, 1990, Section 8.5). The statistic measures (in the previous regression) how much the predicted value of the dependent variable (trip rates) is affected at a particular point (observation) if that observation is deleted. It can be proved that DFFITS is functionally related to the Studentized residuals. If the latter increases, then DFFITS increases too. Both Studentized residuals and DFFITS measures are available in the regression output of software packages for statistical analysis.

Using both measures above, we found a list of unusual cases that is shown in Table 6.3. Case 51, for example, shows a household with one member and no employees taking 24 trips when the average trip rate for the group is 6.28 (Table 6.1). The Studentized residual for this case is 3.93, substantially higher than a cutoff point of 2. As an aside, a cutoff point of 2 would mark 5 percent of the observations as unusual because, under normality assumptions, Studentized residuals follow approximately a *t* distribution.

We also applied the criterion $2[(14 + 1)/360]^{1/2} = 0.408$ as a cutoff point for DFFITS (Sen and Srivavstava, 1990) given 360 observations and 14 independent

Table 6.3 Identified outliers and influential observations

Case number	Studentized residual	DFFITS value	Number of employees	Household size	Number of samples	Number of trips	Average trip rate in group
51	3.93	0.55	0	1	53	24	6.28
91	2.41	0.26	1	1	87	17	6.09
186	3.01	0.46	1	2	45	24	10.42
263	2.32	0.34	2	2	46	20	9.56
273	−2.15	−1.53	0	3	3	8	16.00
275	2.50	0.61	1	3	18	22	10.94
281	2.72	0.66	1	3	18	23	10.94
288	−2.02	−0.49	1	3	18	2	10.94
291	2.27	0.55	1	3	18	21	10.94
300	2.57	0.82	2	3	11	22	10.81
304	−2.48	−2.50	3	3	2	3	11.00
305	2.48	2.50	3	3	2	19	11.00
327	−2.56	−0.49	2	4+	28	4	15.46
340	2.80	0.54	2	4+	28	28	15.46
345	2.58	0.50	2	4+	28	27	15.46
349	6.61	1.35	2	4+	28	45	15.46
350	2.35	0.45	2	4+	28	26	15.46
352	2.44	1.10	3	4+	6	22	11.83

variables (the indicator variables corresponding to each trip rate category in Table 6.1). Not surprisingly, nearly all of the cases with large residuals also had large DFFITS values. An inordinate amount of influence is seen in Table 6.3 in cases stemming from small samples (for example, cases 273, 304, 305, and 352). Case 349 also appears to be quite unusual and influential.

The unusual cases in Table 6.3 normally invite further scrutiny. It is important to understand the dynamics of household formation that give rise to such trip patterns. For example, case 273 belongs to a category (cell) in which two of the three households surveyed were retired families with high incomes, and the third one was a family with several children headed by a student with a part-time job. In another example, case 349, one of the four household members recorded a rather high number of 20 short trips.

The analyst at this point needs to understand what behavior is truly unusual and influential, or simply the result of a small-scale survey. If the former is true, one can drop such observations from further analysis because there is enough evidence that such observations do not belong in the same model. If the latter is true, then a course of action discussed in the next section would improve the reliability of trip rates.

6.5 IMPROVING THE RELIABILITY OF TRIP GENERATION RATES WITH CART PROCEDURES

To improve the reliability of trip rates in the presence of small samples, we propose the use of a CART analysis. CART analysis is based on the binary decision tree algorithm (Breiman et al., 1984). The method is also documented in Ripley (1996) and Venables and Ripley (1997). In this section, the method is explained using two- and three-independent variables examples.

6.5.1 Background Information

As a nonparametric method of estimating conditional distributions, CART models have some potential advantages over parametric models (Reiter, 2003). First, CART modeling may be more easily applied than parametric modeling, particularly for continuous data that are truncated or not smooth. Second, CART models can capture nonlinear relationships and interaction effects that may not be easily revealed in the process of fitting parametric models. Third, CART provides a semi-automatic way to fit the most important relationships in the data, which can be a substantial advantage when there are many potential predictors. Primary disadvantages of CART models relative to parametric models include difficulty of interpretation, discontinuity at partition boundaries, and decreased effectiveness when the data follow relationships easily captured by parametric models (Friedman, 1991).

The CART algorithm is a binary (nodes are always split into two) recursive partitioning algorithm. The original version uses the Gini index of diversity as the default splitting criterion (Breiman et al., 1984). We used Clark and Pregibon's (1992) variation implemented in the *S* language that uses the deviance as the splitting criterion. Clark and Pregibon (1992) define the deviance of a node as the sum of the deviances of all observations in the node (Equation (6.1)). This method is based on an impurity index known as entropy (Ripley, 1996). If the deviance of a node is not zero, and there are sufficient observations in the node, splitting proceeds by comparing the deviance of the node to that of two possible subnodes. The split that maximizes the change in deviance is chosen:

$$D = -2\sum_{j=1}^{J}(n_{jL}\log p_{jL} + n_{jR}\log p_{jR}) \tag{6.1}$$

where D is the deviance, n_{jL} (n_{jR}) denote the number, and p_{jL} (p_{jR}) the proportion of observations in the left (right) node in level J.

CART-type analysis has been growing in popularity because a variety of algorithmic approaches have been implemented in major statistical packages. Some implementations are licensed as add-ons, requiring additional license fees that can be substantial during the licensing period. Many other implementations are disseminated as free software and give practitioners the best chance to start experimenting with these methods. Examples of free software available under the Open Source license agreement include the GUIDE Regression Tree software (http://www.stat.wisc.edu/~loh/guide.html), the WinMine Toolkit (http://research.microsoft.com/%7Edmax/WinMine/tooldoc.htm) and Classification Tree in Excel (http://www.geocities.com/adotsaha/CTree/CtreeinExcel.html), assuming Microsoft Excel is installed. Additional open-source packages available from the Comprehensive R Archive Network (CRAN) include *rpart* (Recursive PARTitioning and regression trees) and *tree* (Classification and regression trees).

6.5.2 CART Demonstration Using Two Independent Variables

Study the CART analysis with the trip generation rates in Table 6.1. Running the same data through the CART algorithm produces the classification shown in Table 6.4. Note that worker and household size categories (0, 3 and 1, 3) are now one category, with 21 households and 245 trips. The sum of the two categories it replaced is shown in Table 6.1. Similarly, categories 2, 3 and 3, 3 have formed a new category, with 13 households and 141 trips. The category formed by categories 0,

Table 6.4 Average trip rate per household after CART classification

<table>
<tr><th rowspan="2">Workers</th><th colspan="4">Household size</th></tr>
<tr><th>1</th><th>2</th><th>3</th><th>4+</th></tr>
<tr><td>0
Number of households</td><td>6.28
53</td><td>9.22
40</td><td rowspan="2">11.67
21</td><td rowspan="2">8.89
18</td></tr>
<tr><td>1
Number of households</td><td>6.09
87</td><td>10.42
45</td></tr>
<tr><td>2
Number of households</td><td>—
—</td><td>9.56
46</td><td rowspan="2">10.85
13</td><td>15.46
28</td></tr>
<tr><td>3
Number of households</td><td>—
—</td><td>—
—</td><td rowspan="2">11.44
9</td></tr>
<tr><td>4
Number of households</td><td>—
—</td><td>—
—</td><td>—
—</td></tr>
</table>

—: indicates category not possible in this classification.

4+ and 1, 4+ has 18 households and 160 trips, and categories 3, 4+ and 4, 4+ have been combined into one category with 9 households and 103 trips.

It is not uncommon to estimate trip production rates by traffic analysis zone (TAZ) using estimated trip generation rates from a household travel survey based on a cross-classification of trip purpose by household size. As an aside, a TAZ is a special area delineated by state and/or local transportation officials for tabulating traffic-related data—especially journey-to-work and place-of-work statistics. A TAZ usually consists of one or more census blocks, block groups, or census tracts (U.S. Census Bureau, 2001).

For example, for each TAZ, CUUATS uses the formula:

$$T_{pk} = \sum_{i} x_{ik} t_{ik} t_{ipk} \tag{6.2}$$

where T_{pk} is the total number of trips by purpose, p in TAZ k, x_{ik} is the number of households in TAZ k in (household size) category i; t_{ik} is the trip rate for households in category i in TAZ k, and t_{ipk} is the percent of trips for households in TAZ k in category i and trip purpose p (CUUATS, Long Range Transportation Plan—Appendix 3: Transportation Model Report, 2004). The trip rates for this particular classification are given in Table 6.5.

The number of cases that contribute to a particular trip purpose and household size combination is shown under the respective trip rate (generally, households contribute to multipurpose trips). For this particular demonstration we have removed trips with either end outside the Champaign–Urbana urbanized area.

An alternative way to compute trip rates by household size and trip purpose is now proposed using the CART method. A CART analysis of Table 6.5 produced

Table 6.5 Trip rates by household size and trip purpose

<table>
<tr><th rowspan="2">Trip purpose</th><th colspan="4">Household size</th></tr>
<tr><th>1</th><th>2</th><th>3</th><th>4+</th></tr>
<tr><td>Home-based work
Number of cases</td><td>1.90
71</td><td>2.43
70</td><td>2.60
25</td><td>2.64
36</td></tr>
<tr><td>Home-based shop
Number of cases</td><td>1.64
42</td><td>2.26
76</td><td>2.50
20</td><td>2.81
32</td></tr>
<tr><td>Home-based school
Number of cases</td><td>2.19
58</td><td>2.26
23</td><td>1.22
9</td><td>2.72
11</td></tr>
<tr><td>Home-based other
Number of cases</td><td>2.44
93</td><td>4.00
105</td><td>3.97
29</td><td>6.14
47</td></tr>
<tr><td>Non home-based
Number of cases</td><td>2.82
99</td><td>3.61
107</td><td>4.48
29</td><td>4.61
41</td></tr>
</table>

Table 6.6 Trip rates by household size and trip purpose (after CART analysis)

<table>
<tr><th rowspan="2">Trip purpose</th><th colspan="4">Household size</th></tr>
<tr><th>1</th><th>2</th><th>3</th><th>4+</th></tr>
<tr><td>Home-based work
Number of cases</td><td rowspan="3">1.94
171</td><td rowspan="3" colspan="2">2.33
223</td><td rowspan="3">2.72
79</td></tr>
<tr><td>Home-based shop
Number of cases</td></tr>
<tr><td>Home-based school
Number of cases</td></tr>
<tr><td>Home-based other
Number of cases</td><td>2.44
93</td><td>4.00
105</td><td rowspan="2">4.22
58</td><td>6.15
47</td></tr>
<tr><td>Non home-based
Number of cases</td><td>2.82
99</td><td>3.61
107</td><td>4.61
41</td></tr>
</table>

the trip rates shown in Table 6.6. The top number in each category is the trip rate, and the bottom number is the number of households in the same category. The previous 20 categories in Table 6.5 have been fused into 10 categories in Table 6.6 with much larger sample sizes (it can readily be verified that, notwithstanding rounding errors, Tables 6.5 and 6.6 give the same total number of trips). In this regard, Table 6.6 is a reasonable alternative to Table 6.5 and offers the additional advantage that the estimation of trip rates are based on more samples.

The results from the CART analysis in Table 6.6 show that small-size (single-person), medium-size (two- and three-person), and larger-size (four-or-more person) households appear to have distinctive trip-making profiles in regard to relatively less discretionary (or more mandatory) trips (home-based work, home-based shop, and home-based school) as opposed to relatively more discretionary trips (home-based other and non home-based). If this observation can be validated from other reliable sources, it could have important implications in the design of household travel surveys.

6.5.3 Comparison Between CUUATS and CART Models

Using the trip rates from Table 6.6, we can estimate the total number of trips from the formula:

$$T_{pk}^{CART} = \sum_i x_{ik}^{CART} t_{ipk}^{CART} \tag{6.3}$$

where T_{pk}^{CART} is the total number of trips in TAZ k by purpose p using the CART procedure, x_{ik} is the number of households and trip purpose combinations in (household size) category i and TAZ k as determined by the CART procedure; and t_{ipk}^{CART} is the trip rate for households in category i and trip purpose p in TAZ k, also determined by the CART procedure.

Clearly, the models in Equations (6.2) and (6.3) are different algebraically. However, they both give similar total number of trips by purpose as shown in Table 6.7. As a result, the CART approach appears to be a reasonable alternative to more traditional procedures for trip generation estimation. The added value for the CART approach, however, is that it is based on trip rates that have been estimated from a richer sample of households.

6.5.4 CART Demonstration Using Three Independent Variables

Using the same data and considering trip purpose as an additional independent variable, Table 6.8 presents average trip rates per household for trips internal to the study area. According to the survey documentation (Morocoima-Black and Kang, 2003), home-based work trips include trips from home to work, work-related business, and return home. Home-based shopping trips include any kind of trips for shopping and the return home. Home-based school trips are from home to school and back home. Home-based other is a category for the remainder of home-based trips. Non home-based trips are those that do not begin or end at home.

A regression tree for the previous three-way classification is shown in Table 6.9. The response variable is the number of trips. The independent variables are number of workers per household (five categories), household size (four categories), and trip purpose (five categories). Of the 100 possible category combinations, only 70 are possible; however, five categories have missing observations on the trip purpose variable. Thus only 65 categories are included in the analysis.

A re-expression of the results in Table 6.9 is shown in Table 6.10. The number in each category combination is the trip rate corresponding to the respective regression tree terminal node (at which point further group splitting stops) in Table 6.9.

Table 6.7 Number of trips comparison between CUUATS and CART models

Trip purpose	1-person households		2-person households		3-person households		4+ person households		Total number of trips in survey	
	CUUATS	CART	CUUATS	CART	CUUATS	CART	CUUATS	CART	CUUATS	CART
Home-based work	137.7	136.6	163.1	166.9	58.3	58.1	97.9	97.8	457.0	459.5
Home-based shopping	81.5	76.9	177.1	179.7	46.6	46.5	87.0	83.8	392.2	386.9
Home-based school	112.5	111.0	53.6	51.4	21.0	19.4	29.9	27.9	217.0	209.7
Home-based other	226.9	230.6	420.0	423.7	122.4	120.2	289.1	286.4	1058.4	1060.8
Non home-based	280.2	281.8	387.3	385.1	122.4	124.0	189.0	195.6	978.9	986.6
Total number of trips	838.8	836.9	1201.1	1206.8	370.6	368.2	692.9	691.5	3103.5	3103.4

Table 6.8 Mean trip rates for a three-way classification

Workers	Household size	Trip purpose	Number of households	Mean trip rate	Variance	St. error
0	1	hb-work	8	1.50	0.29	0.19
	1	hb-shop	17	1.65	0.99	0.24
	1	hb-school	28	2.29	1.47	0.23
	1	hb-other	39	2.64	3.08	0.28
	1	nh-based	39	3.10	5.20	0.37
0	2	hb-work	11	1.91	0.89	0.28
	2	hb-shop	25	2.44	2.34	0.31
	2	hb-school	8	2.25	1.36	0.41
	2	hb-other	34	4.24	4.97	0.38
	2	nh-based	32	3.16	3.36	0.32
0	3	hb-work	1	2.00	ND**	ND
	3	hb-shop	2*	5.50	0.50	0.50
	3	hb-school	1	1.00	ND	ND
	3	hb-other	3	6.00	7.00	1.53
	3	nh-based	3	4.00	1.00	0.58
0	4	hb-work	0	ND	ND	ND
	4	hb-shop	1	1.00	ND	ND
	4	hb-school	2	2.00	2.00	1.00
	4	hb-other	3	4.33	2.33	0.88
	4	nh-based	3*	2.00	0.00	0.00
1	1	hb-work	63	1.95	0.88	0.12
	1	hb-shop	25	1.64	1.24	0.22
	1	hb-school	30	2.10	1.33	0.21

	1	hb-other	54	2.30	2.85	0.23
	1	nh-based	60	2.65	2.77	0.22
1	2	hb-work	26	2.04	0.68	0.16
	2	hb-shop	23	2.04	1.04	0.21
	2	hb-school	10	2.20	0.84	0.29
	2	hb-other	36	4.17	7.00	0.44
	2	nh-based	36	4.56	11.11	0.56
1	3	hb-work	14	2.07	0.69	0.22
	3	hb-shop	12	2.00	1.45	0.35
	3	hb-school	7	1.14	0.14	0.14
	3	hb-other	15	3.67	8.10	0.73
	3	nh-based	16	5.00	17.60	1.05
1	4	hb-work	9	2.56	1.28	0.38
	4	hb-shop	5	2.60	3.80	0.87
	4	hb-school	3*	2.00	0.00	0.00
	4	hb-other	13	5.54	6.94	0.73
	4	nh-based	8	3.25	3.93	0.7
2	2	hb-work	33	2.91	3.40	0.32
	2	hb-shop	28	2.29	1.25	0.21
	2	hb-school	5	2.40	1.30	0.51
	2	hb-other	35	3.60	5.42	0.39
	2	nh-based	39	3.13	2.69	0.26
2	3	hb-work	8	3.13	4.70	0.77
	3	hb-shop	6	2.50	1.10	0.43
	3	hb-school	1	2.00	ND	ND
	3	hb-other	9	3.78	5.94	0.81

Table 6.8 *Continued*

Workers	Household size	Trip purpose	Number of households	Mean trip rate	Variance	St. error
	3	nh-based	8	4.38	7.13	0.94
2	4	hb-work	18	2.78	3.12	0.42
	4	hb-shop	20	3.20	3.85	0.44
	4	hb-school	6	3.33	6.27	1.02
	4	hb-other	24	6.71	25.35	1.03
	4	nh-based	24	5.71	15.87	0.81
3	3	hb-work	2*	4.50	24.50	3.50
	3	hb-shop	0	ND	ND	ND
	3	hb-school	0	ND	ND	ND
	3	hb-other	2	4.00	18.00	3.00
	3	nh-based	2	1.50	0.50	0.50
3	4	hb-work	6	2.50	1.50	0.50
	4	hb-shop	4	2.25	1.58	0.63
	4	hb-school	0	ND	ND	ND
	4	hb-other	5	7.20	27.70	2.35
	4	nh-based	4	2.75	2.25	0.75
4	4	hb-work	3	2.33	2.33	0.88
	4	hb-shop	2	1.50	0.50	0.50
	4	hb-school	0	ND	ND	ND
	4	hb-other	2*	3.50	0.50	0.50
	4	nh-based	2	4.50	12.5	2.50

*standout categories; ND: no data.

Table 6.9 Regression tree for the three-way classification

Regression tree terminal node	Number of workers per household categories	Household size categories	Trip purpose* categories	Number of groups in terminal node**	Number of households	Mean Trip rate per household
1	0, 1, 4	1, 2, 3, 4+	1, 2, 3	336	235	2.91
2	2, 3	1, 2, 3, 4+	1, 2, 3	137	89	4.28
3	0, 1	1	4, 5	192	119	4.26
4	0, 2, 3	2, 3	5	84	84	3.25
5	0, 2, 3	2, 3	4	83	83	3.98
6	1	2, 3	4, 5	103	58	7.74
7	0, 1, 3, 4	4+	5	17	17	3.06
8	0, 1, 4	4+	4	18	18	5.11
9	3	4+	4	5	5	7.20
10	2	4+	4, 5	48	26	11.46

Residual mean deviance = 4.38. *Trip purpose: 1=home-based work; 2=home-based shopping; 3=home-based school; 4=home-based other; 5=non home-based. **The term *group* is meant to denote category combinations.

Table 6.10 Trip rates by household size, number of workers and trip purpose (after CART analysis)

<table>
<tr><th rowspan="2">Workers</th><th rowspan="2">Trip purpose</th><th colspan="4">Household size</th></tr>
<tr><th>1</th><th>2</th><th>3</th><th>4+</th></tr>
<tr><td>0</td><td>hb-work</td><td colspan="4" rowspan="3">2.91</td></tr>
<tr><td></td><td>hb-shop</td></tr>
<tr><td></td><td>hb-school</td></tr>
<tr><td></td><td>hb-other</td><td rowspan="2">4.26</td><td colspan="2">3.98</td><td>5.11</td></tr>
<tr><td></td><td>nh-based</td><td colspan="2">3.25</td><td>3.06</td></tr>
<tr><td>1</td><td>hb-work</td><td colspan="4" rowspan="3">2.91</td></tr>
<tr><td></td><td>hb-shop</td></tr>
<tr><td></td><td>hb-school</td></tr>
<tr><td></td><td>hb-other</td><td rowspan="2">4.26</td><td colspan="2" rowspan="2">7.74</td><td>5.11</td></tr>
<tr><td></td><td>nh-based</td><td>3.06</td></tr>
<tr><td>2</td><td>hb-work</td><td>—</td><td colspan="3" rowspan="3">4.28</td></tr>
<tr><td></td><td>hb-shop</td><td>—</td></tr>
<tr><td></td><td>hb-school</td><td>—</td></tr>
<tr><td></td><td>hb-other</td><td>—</td><td colspan="2">3.98</td><td rowspan="2">11.46</td></tr>
<tr><td></td><td>nh-based</td><td>—</td><td colspan="2">3.25</td></tr>
<tr><td>3</td><td>hb-work</td><td>—</td><td>—</td><td colspan="2" rowspan="3">4.28</td></tr>
<tr><td></td><td>hb-shop</td><td>—</td><td>—</td></tr>
<tr><td></td><td>hb-school</td><td>—</td><td>—</td></tr>
<tr><td></td><td>hb-other</td><td>—</td><td>—</td><td>3.98</td><td>7.20</td></tr>
<tr><td></td><td>nh-based</td><td>—</td><td>—</td><td>3.25</td><td>3.06</td></tr>
<tr><td>4</td><td>hb-work</td><td>—</td><td>—</td><td>—</td><td rowspan="3">2.91</td></tr>
<tr><td></td><td>hb-shop</td><td>—</td><td>—</td><td>—</td></tr>
<tr><td></td><td>hb-school</td><td>—</td><td>—</td><td>—</td></tr>
<tr><td></td><td>hb-other</td><td>—</td><td>—</td><td>—</td><td>5.11</td></tr>
<tr><td></td><td>nh-based</td><td>—</td><td>—</td><td>—</td><td>3.06</td></tr>
</table>

—: indicates category not possible in this classification.

A common strategy for building trees is to fit one with a large numbers of nodes and then prune the tree according to some optimality or complexity criteria. Pruned trees typically do not predict the values in the observed data as well as larger ones, but they can be more robust to overfitting than larger ones. The CART algorithm treats pruning as a tradeoff between two issues: getting the right size tree and getting accurate estimates of the true probabilities of misclassification. This process is known as minimal cost complexity pruning.

For moderate size samples (on the order of 1000), the above method can be used in combination with cross validation (Lachenbruch and Mickey, 1968). The idea is that, instead of using one sample (training data) to build a tree and another sample (pruning data) to test the tree, one can form several pseudo-independent samples from the original sample and use these to form a more accurate estimate of the classification error. A ten-fold cross-validation is recommended for the CART algorithm. This is accomplished by holding out 10 percent of the data, fitting a tree to the other 90 percent of the data, and dropping the held-out data through the tree. While doing so, we note at which level the tree gives the best results. Then we hold out a different 10 percent and repeat.

The fully-grown tree of the previous three-way classification would have 40 terminal nodes. After pruning, the remaining portion of the tree would have 10-12 terminal nodes, accounting for about 50 percent of the total variation in the number of trips (Figure 6.1). The tree with 10 terminal nodes is shown in Tables 6.9 and 6.10. Minimal gain in residual deviance would have resulted had we chosen to retain more terminal nodes by continuing group splitting.

It is obvious from Tables 6.9 and 6.10 that home-based work, shopping, and school trips cluster differently from home-based other and non home-based trips, independent of family size. Perhaps this is not unexpected because the second group is comprised of trips seemingly more discretionary in nature than the first

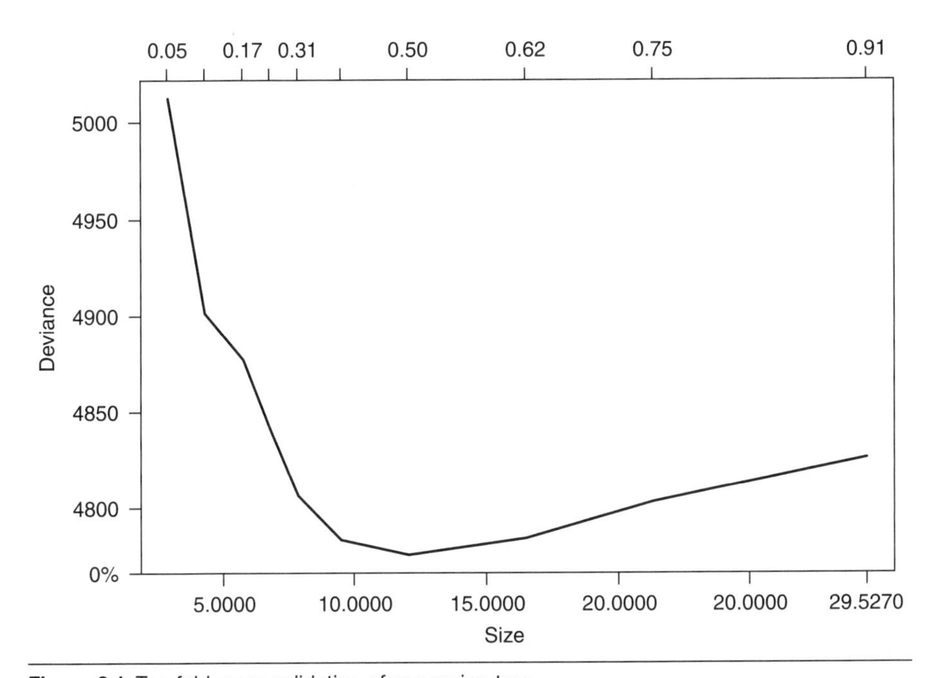

Figure 6.1 Ten-fold cross validation of regression tree

group. In addition, single-member households are making discretionary trips at different rates than two- and three-member families, and larger families as well.

In the future, it would be interesting to investigate whether these observations reflect local travel behavior or can be corroborated in larger, perhaps national-scale surveys. Clearly, if the former is true, it could limit the potential for transferability of household trip generation rates. On the contrary, if the latter proves to be true, then what a relief for the local transportation planners!

In numerous studies, vehicle availability has been reported to be an important predictor in trip generation. Intrigued by the small samples in Table 6.2, we decided to conduct a CART analysis to obtain more reliable trip generation rates for this particular three-way classification. The 2002 travel survey (Morocoima-Black and Kang, 2003) collected data on vehicles available for use by members of the household. Four levels of vehicle availability are reported: none, one, two, and three or more vehicles in the household.

Results from a CART analysis of Table 6.2 are shown in Tables 6.11 and 6.12. The number in each category combination of Table 6.12 is the trip rate for each category combination corresponding to the respective regression tree terminal node in Table 6.11. Tree growing and pruning followed the same rules as previously discussed. The 10 terminal nodes in Tables 6.11 and 6.12 captured more than 70 percent of the total deviance. Again, little additional deviance reduction was achieved with additional tree growing.

Table 6.11 Regression tree analysis of Table 6.2

Regression tree terminal node	Number of workers per household categories	Household size categories	Vehicle availability categories	Number of households	Mean trip rate per household
1	0, 1	1	0, 1, 2, 3+	140	6.16
2	0	2	0, 1	14	10.14
3	0	2	2, 3+	26	8.73
4	1, 2	2	0, 1, 2, 3+	91	9.89
5	0, 1	3, 4+	0, 1	23	9.09
6	0, 1	3	2, 3+	9	14.89
7	0, 1	4+	2, 3+	7	8.86
8	2, 3	3	0, 1, 2, 3+	13	10.85
9	2	4+	0, 1, 2, 3+	28	15.46
10	3, 4	4+	0, 1, 2, 3+	9	11.44

Table 6.12 Re-expression of Table 6.2 (after CART analysis)

<table>
<tr><th rowspan="2">Workers</th><th rowspan="2">Vehicle availability</th><th colspan="4">Household size</th></tr>
<tr><th>1</th><th>2</th><th>3</th><th>4+</th></tr>
<tr><td>0</td><td>0</td><td rowspan="8">6.16</td><td rowspan="2">10.14</td><td colspan="2" rowspan="2">9.09</td></tr>
<tr><td></td><td>1</td></tr>
<tr><td></td><td>2</td><td rowspan="2">8.73</td><td rowspan="2">14.89</td><td rowspan="2">8.86</td></tr>
<tr><td></td><td>3+</td></tr>
<tr><td>1</td><td>0</td><td rowspan="8">9.89</td><td colspan="2" rowspan="2">9.09</td></tr>
<tr><td></td><td>1</td></tr>
<tr><td></td><td>2</td><td rowspan="2">14.89</td><td rowspan="2">8.86</td></tr>
<tr><td></td><td>3+</td></tr>
<tr><td>2</td><td>0</td><td>—</td><td rowspan="8">10.85</td><td rowspan="4">15.46</td></tr>
<tr><td></td><td>1</td><td>—</td></tr>
<tr><td></td><td>2</td><td>—</td></tr>
<tr><td></td><td>3+</td><td>—</td></tr>
<tr><td>3</td><td>0</td><td>—</td><td>—</td><td rowspan="8">11.44</td></tr>
<tr><td></td><td>1</td><td>—</td><td>—</td></tr>
<tr><td></td><td>2</td><td>—</td><td>—</td></tr>
<tr><td></td><td>3+</td><td>—</td><td>—</td></tr>
<tr><td>4</td><td>0</td><td>—</td><td>—</td><td>—</td></tr>
<tr><td></td><td>1</td><td>—</td><td>—</td><td>—</td></tr>
<tr><td></td><td>2</td><td>—</td><td>—</td><td>—</td></tr>
<tr><td></td><td>3+</td><td>—</td><td>—</td><td>—</td></tr>
</table>

—: indicates category not possible in this classification.

Tables 6.11 and 6.12 show that vehicle availability influences the trip making behavior of two-member and larger households. In those cases, the trip rates are different for one-car or fewer households as compared to multiple-car households. This is an important finding because it is uncommon in traditional trip generation models to uncover such distinctive behavior.

From a practical standpoint, it is interesting to contrast Tables 6.2, 6.11, and 6.12. Table 6.2 shows 56 category combinations of which 18 (flagged with ND in Table 6.2) had no samples. After the CART analysis, each category combination has at least seven households from which to draw trip rates. This is one way to boost small-sample household travel survey reliability.

6.6 ROW-COLUMN DECOMPOSITION ANALYSIS AS AN IMPUTATION METHOD

The third problem addressed is the issue of missing (no samples) or minimal (few samples) information in small-scale household travel surveys. We propose to ameliorate such issues using row-column decomposition analysis as an imputation method.

Row-column decomposition analysis is a simple procedure for the analysis of travel data, and its implementation has become incidental with today's spreadsheet functionality. The method has been utilized for trip generation (see elementary analysis in Sen and Johnson, 1977), trip distribution (see elementary analysis in Sööt and Sen, 1978), and mode split (see elementary analysis in Sen and Johnson, 1977). Tukey (1977) has discussed such procedures under the name of row-PLUS-column fit.

Row-column decomposition analysis of Table 6.1 is carried out in the two steps shown in Tables 6.13 and 6.14, respectively. Table 6.13 shows the first step of a row-column decomposition analysis. The mean for each column has been computed and subtracted from the original cell values of Table 6.1. The means themselves, called column fits, are written at the bottom of Table 6.13.

Table 6.14 completes the row-column decomposition analysis started in Table 6.13. Now the means of the rows in Table 6.13 are subtracted, the means themselves noted in the right margin as row effects. The grand mean is computed as the mean of the column fits; the values of the column fit less than the grand mean are shown in the bottom row of Table 6.14 as column effects. The numbers in the left in the cells of the cross tabulations in Table 6.14 are the residuals.

Table 6.14 can be used to reconstruct Table 6.1 by using the formula:

$$y_{ij} = \mu + a_i + b_j + r_{ij} \tag{6.4}$$

where y_{ij} is the original observation in the ith row and jth column of Table 6.1, μ is the grand mean, a_i is the effect of the ith row, b_j is the effect of the jth column, and r_{ij} is the residual in the ith row and jth column.

Any decomposition of the s in the form of Equation (6.4) is a row-column decomposition analysis (Sen and Johnson, 1977). Tukey (1977) also discusses further details on various forms of row-column decomposition analysis and the types of transformations that may be required to achieve the type of additivity implied by Equation (6.4). Sen and Johnson (1977) argue that row-column decomposition analysis is a regression technique and that row-column decomposition analysis by means is a least-squares technique. The method does not lend itself to formal hypothesis testing, however, as a method for simple model building, it compares favorably with traditional methods used in the analysis of travel data: regression, category analysis, and analysis of variance.

Table 6.13 Step 1: Column means subtracted

Workers	Household size			
	1	2	3	4+
0	0.10	−0.51	3.81	−3.00
1	−0.10	0.69	−1.25	−1.94
2	U*	−0.17	−1.38	4.46
3	U	U	−1.19	0.83
4	U	U	U	−0.34
Column fit	6.18	9.73	12.19	11.00

*U - unavailable

Examples: Column fit = means of column; for example, 6.19 = (6.28+6.09)/2 (from Table 6.1) Cell value = observed value − column fit; for example, 0.10 = 6.28 − 6.18

Table 6.14 Step 2: Row means subtracted

Workers	Household size				Row effects
	1	2	3	4+	
0	−0.00	−0.61	3.71	−3.10	0.10
1	0.55	1.34	−0.60	−1.29	−0.65
2	U	−1.14	−2.35	3.49	0.97
3	U	U	−1.01	1.01	−0.18
4	U	U	U	0	−0.34
Column effects	−3.60	−0.04	2.41	1.23	Grand mean 9.78

Examples: Row effect = mean of step one row; for example, 0.10 = (0.10 − 0.51 + 3.81 − 3.00)/4
Residuals = cell value of step one − row effect; for example, −0.00 = 0.10 − 0.10
Grand mean = mean of column fits; for example, 9.78 = (6.18 + 9.73 + 12.19 + 11.00)/4
Column effect = column fit − grand mean; for example, −3.60 = 6.18 − 9.78
Original cell value = grand mean + row effect + column effect + residual; for example, 6.28 = 9.78 + 0.10 − 3.60 − 0.00

As shown in Table 6.14 (number of workers and household size) categories (0, 3) and (0, 4+) with three observations, each (Table 6.1) has high residuals. In category (0, 3), two of the three households were retired families with high incomes, and the third was a family with several children headed by a student with a part-time job. In the (0, 4+) category, of the three families surveyed, one was a retired high-income family and two were families with several children headed by an unemployed international student.

Category (2, 4+) also has a high residual value as corroborated in Table 6.3 in which several cases in this category were identified as unusual. In this category we found families with two children who are either overly or minimally active regarding trip making. Such cases would need further scrutiny as discussed earlier.

We repeated the analysis several times after removing the unusual cases reported in Table 6.3 and observed some improvement regarding the size of the residuals—but not at a desired level. Perhaps some level of re-expression for the dependent variable (trip rate) is needed to achieve the type of additivity implied by Equation (6.4) as Tukey (1977) suggested. Note that all independent variables in this case are categoric (0/1 variables), and any transformation is meaningless for them. After a logarithmic transformation of the trip rates in Table 6.1 (without removing any unusual observations), a row-column decomposition analysis was performed, as shown in Table 6.15.

The residuals of the logarithmic transformation in Table 6.15 show a substantially better fit than the raw values would yield. Notice that once the rather strong effect of household size has been accounted for, the effect of the number of workers is minimal. Since we transformed the dependent variable (trip rate) it would be prudent to recheck heteroscedasticity, perhaps continue the search for a better transformation, and repeat the previous steps.

Row-column decomposition analysis can also be applied to *n*-way tables. For example, a row-column decomposition analysis of Table 6.2 would treat the three-way classification (number of workers, family size, and auto availability) as a two-way table (for example, [number of workers, auto availability], family size) and proceed with the steps described. Similarly, a row-column decomposition analysis of a four-way table could be translated into a two-way *analysis* using the category combinations of three of the four variables in rows. Additional dimensions could be accommodated in an analogous fashion.

To our knowledge, imputation in trip generation is either uncommon or seldom discussed. As we have seen, opportunities for imputation arise naturally in trip generation modeling and techniques developed for missing data, and contingency table imputation could find fertile ground here. For example, Brownstone (1998) showed how Rubin's (1987) multiple imputation methodology could help alleviate problems caused by survey nonresponse and missing data. Wang and Shao (2003) used hot-deck imputation for missing data in a two-way contingency table, whereas Cox (2002) used linear programming and MCMC methods for the same problem.

On the other hand, research into the use of CART-type methods for imputation has not been conclusive, and other algorithmic approaches appear to be more promising (Michie et al., 1994). Reiter (2003) reports that sequential CART models have promise as a method for generating partially synthetic data sets. He also reports that the primary drawback of the approach is the sequential nature

Table 6.15 Row-column decomposition analysis (logarithmic transformation of trip rates in Table 6.1)

Workers	Household size				
	1	2	3	4+	
0	1.84	2.22	2.77	2.08	
1	1.81	2.34	2.39	2.20	
2	U*	2.26	2.38	2.74	
3	U	U	2.40	2.47	
4	U	U	U	2.37	
Step 1: Column means subtracted					
Workers	**Household size**				
	1	**2**	**3**	**4+**	
0	0.02	−0.05	0.29	−0.29	
1	−0.02	0.07	−0.09	−0.17	
2	U	−0.02	−0.11	0.37	
3	U	U	−0.09	0.10	
4	U	U	U	−0.01	
Column fit	1.82	2.27	2.49	2.37	
Step 2: Row means subtracted					
Workers	**Household size**				**Row effects**
	1	**2**	**3**	**4+**	
0	0.03	−0.04	0.30	−0.28	−0.01
1	0.04	0.12	−0.04	−0.12	−0.05
2	U	−0.10	−0.19	0.29	0.08
3	U	U	0.09	0.09	0.01
4	U	U	U	0.00	−0.01
Column effects	−0.42	0.04	0.25	0.13	Grand mean 2.24

*U—unavailable.

of the imputations, which can introduce conditional independence structures into the released data. This issue also affects the use of CART models, or any sequential imputation scheme, for imputation of missing data.

In addition, Wilmot and Shivananjappa (2003) tested the accuracy of CART-type and neural-network imputation procedures on a sample of households from the 1995 Nationwide Personal Transportation Survey. The analysis produced mixed results for the variables tested (household income, number of vehicles, educational status, and age), and Breiman (2003) discussed imputation using random forests.

Meanwhile, the simplicity in implementation of the model in Equation (6.4) shows that row-column decomposition analysis can be used for imputation in cases in which one or more independent variables have missing information or in cases in which the number of observations is small. Indeed, once the residuals have been minimized, and a satisfactory model has been obtained, Equation (6.4) can be used to impute the missing cell value by the value of the grand mean augmented by the values of the respective row and column effects. We believe that the method compares favorably (in terms of cost effectiveness) with other more complicated imputation approaches and could become a valuable tool especially among transportation modelers in smaller MPOs.

6.7 SUMMARY

The analyst of small-scale household travel surveys is facing a number of issues when estimating or validating trip generation rates as an input to the trip generation modeling process. Such issues include data quality screening, reliability of trip rates in the presence of small samples, and imputation issues. This chapter has proposed methods to tackle each of these tasks. The methods are fairly easy to implement in a cost-effective manner.

REFERENCES

Breiman, L., J. H. Friedman, R. A. Olshen, and C. J. Stone. 1984. *Classification and regression trees (CART)*. Pacific Grove, CA: Wadsworth.

Breiman, L. 2003. Manual for setting up, using, and understanding random forest V4.0. http://oz.berkeley.edu/users/breiman/Using_random_forests_v4.0.pdf.

Brownstone, D. 1998. Multiple imputation methodology for missing data, non-random response and panel attrition. Theoretical foundations of travel choice modeling, 421–449. University of California Transportation Center. http://www.uctc.net/papers/594.pdf.

Clarke, L. A. and D. Pregibon. 1992. Tree-based models. In *Statistical Models in* S, eds. J. M. Chambers and T. J. Hastie, 377–419. Pacific Grove, CA: Wadsworth & Brooks/Cole.

Champaign–Urbana urbanized area transportation study. 2004. Long range transportation plan 2025. December 2004.

Cox, L. H. 2002. Imputing missing values in two-way contingency tables using linear programming and Markov Chain Monte Carlo. UNECE Work Session on Statistical Data Editing, Working paper no. 39.

Friedman, J. H. 1991. Multivariate adaptive regression splines (with discussion). *The Annals of Statistics*, 19: 1–141.

Lachenbruch, P. and R. Mickey. 1968. Estimation of error rates in discriminant analysis. *Technometrics*, 10: 1–11.

Michie, D., D. J. Spiegelhalter, and C. C. Taylor, eds. 1994. *Machine learning, neural and statistical classification*. Ellis Horwood Series in Artificial Intelligence. Ellis Horwood.

Morocoima-Black, R. and E. Kang. 2003. 2002 Champaign–Urbana–Savoy travel survey. Final Report. CUUATS.

Reiter, J. P. 2003. Using CART to generate partially synthetic, public use microdata. Proceedings of the Federal Committee on Statistical Methodology Research Conference.

Ripley, B. D. 1996. *Pattern Recognition and Neural Networks*. Cambridge, MA: Cambridge University Press.

Rubin, D. B. 1987. *Multiple Imputation for Nonresponse in Surveys*. New York: John Wiley.

Sen, A. and C. Johnson. 1977. A simple method for analysis of travel data. *Transportation Research*, 11: 189–196.

Sen, A. and M. Srivastava. 1990. *Regression Analysis: Theory, Methods and Applications*. New York: Springer-Verlag.

Sööt, S. and A. Sen. 1978. Elementary analysis in trip distribution modeling. *Transportation Engineering Journal*, ASCE, 104 (TE6). Proc. Paper 14114 (November), 789–797.

Thakuriah, P., A. Sen, S. Sööt, and E. Christopher. 1993. Nonresponse bias and trip generation models. *Transportation Research Record*, 1412: 64–70.

Tukey, J. W. 1977. *Exploratory Data Analysis*. Reading, MA: Addison-Wesley.

Venables, W. N. and B. D. Ripley. 1997. *Modern Applied Statistics with* S-*Plus*. New York: Springer-Verlag.

Wang, H. and J. Shao. 2003. Two-way contingency tables under conditional hot deck imputation. *Statistica Sinica*, 13: 613–623.

Wilmot, C. G. and S. Shivananjappa. 2003. Comparison of hot-deck and neural-network imputation. *Transport Survey Quality and Innovation*, 543–554.

U.S. Census Bureau. 2001. Cartographic boundary files. U.S. Census Bureau, Geography Division, Cartographic Products Management Branch.

7

RECENT PROGRESS ON ACTIVITY-BASED MICROSIMULATION MODELS OF TRAVEL DEMAND AND FUTURE PROSPECTS

Abolfazl Mohammadian
Department of Civil and Materials Engineering
University of Illinois–Chicago

Joshua Auld
Department of Civil and Materials Engineering
University of Illinois–Chicago

Sadayuki Yagi
Japan Research Institute

7.1 INTRODUCTION

Recently, there has been significant progress in the area of activity-based microsimulation used in travel-demand analysis. Activity-based microsimulation models have grown in importance, and criticism of traditional travel demand models has also grown. However, traditional four-step travel demand models continue to make up the majority of the models used in practice. It is hoped that the use of activity-based models will grow with further improvements in model technique, validation, and transferability. This chapter describes some fundamental concepts of activity-based modeling and provides an overview of the statistical procedures used in the model components. Future directions for research and analysis are also proposed.

7.2 BACKGROUND

The traditional approach to estimating travel demand to determine congestion, environmental, and other impacts of the transportation system has been to develop what is known as a four-step travel demand model. In this type of model, travel demand is modeled sequentially with the number of trips from and to each area determined first, followed by the distribution of those trips between all of the areas and the modes and routes chosen for the trips. The traditional four-step model has often been criticized for a variety of shortcomings, including a lack of transferability, poor treatment of time-of-day characteristics, the independence of each trip made by an individual from other trips, and the lack of a behavioral component to trip making due to the use of highly aggregated data (Boyce, 2002; Kitamura, 1996). These models originated in the 1950s and have not changed appreciably since, with the exception of the addition of feedback loops between the traffic assignment results and the cost input to the gravity modeling stage which attempts to simulate congestion effects (Boyce, 2002). The most important limitation of the four-step model in analyzing current transportation demand strategies is its lack of a behavioral focus and the independence of each individual trip. The other deficiencies of the model can be overcome to a certain extent through the use of time-of-day separated Origin-Destination (O-D) matrices, feedback loops, and other strategies. However, the independence of individual trips and the lack of individual behavioral representation present problems that cannot be overcome for analyzing the complex situations currently desired (Kitamura, 1996). An example is a transportation demand strategy that encourages workers to telecommute instead of working at an office. The traditional four-step model would show a reduction in overall trips, because the work trip would be eliminated. However, in actual practice, in-home workers tend to increase their total number of discretionary trips compared to commuters. Similar results occur with programs that encourage transit ridership, wherein trips that were previously completed on the way home from work now generate an entirely separate tour, because the home-work-shopping-home trip tour becomes a home-work-home-shopping tour (Bowman and Ben-Akiva, 1996). In both cases, the use of a transportation demand mitigation strategy caused an overestimation in the reduction of trips in the four-step model because responses such as trip chaining and rescheduling behavior were not considered.

7.3 ACTIVITY-BASED MODELING

Activity-based modeling was proposed to correct the deficiencies observed in the traditional four-step approach. The theory of activity-based modeling states that travel is derived from individual participation in activities and that the participation in activities is driven by fundamental human needs or desires, including basic needs such as food, shelter, clothing, and the more complex requirements of socialization, education, and recreation (Ettema and Timmermans, 1997). The decision to participate in activities, and, therefore travel, although based on basic human needs, can vary due to several factors, including personal characteristics, household composition, and the availability of opportunities to satisfy those needs. Additionally, the ability of an individual to participate in activities is constrained. Several constraints are identified by Hagerstrand in his work on time geography (Hagerstrand, 1970) (Hayes-Roth & Hayes-Roth, 1979). These constraints include capability constraints, referring to the physical limitations on human activity and coupling constraints; an individual's ability to participate in activities is limited by the requirement that they either need to participate in an activity with another person or share a resource with another person at a specific time. Hagerstrand used these constraints to develop a space-time prism that included all feasible activity patterns. Attempts have been made to relate the activity types to broad project categories which cover the basic needs of individuals as an organizing principle (Miller, 2005). This allows provisional schedules for each project—called the project agendas—to be created in advance. Then actual schedules can be created from the provisional schedules within each project, a process that would better represent people's actual scheduling process. The combination of all institutional constraints (store hours, work schedules, and others) with household constraints (joint activities, childcare, and maintenance needs) and resource and space-time constraints determines how people schedule activities to satisfy their needs.

Activity-based modeling has been discussed in the literature since the 1970s, but practical applications have been implemented only recently (Shiftan et al., 2003). Using data from a week-long activity diary, Van der Hoorn (1983) developed an activity-based model, including a set of logit type models for the choice of an activity and a location, given a previous location. Kitamura and Kermanshah (1983) developed a sequential logit-based model of out-of-home activity and destination choice in which the choice of one activity is assumed to be conditioned upon the previous activity. Axhausen (1990) used a discrete event simulation approach to develop a model of activities and network traffic flows for Germany, assuming that activity chains are known and used to schedule event durations. Several other academic studies focused on developing activity-based models in the late 1980s and early 1990s, including STARCHILD (Recker and McNally, 1986), SMASH (Ettema et al., 1993), and SCHEDULER (Golledge et al., 1994). The revolution in hardware and software capabilities and advances in simulation techniques in the late 1990s as well as the advent of the new millennium resulted in several activity-based microsimulation models, including those described in the next section. So far, two general themes or model types have developed within the broader set of activity-based modeling. These include models which use econometric or utility-maximizing approaches to decision behavior and models which attempt to represent the actual cognitive process of decision making in the schedule-creation process (Ettema and Timmermans, 1997). An emerging generation of the activity-based models called *dynamic activity scheduling* is discussed at the end of this chapter.

7.4 ECONOMETRIC ACTIVITY-BASED MODELING

Econometric activity-based models typically use the utility maximization theory to determine the scheduling choices of individuals. These models normally use a logit or nested-logit model to represent the decision-making processes which in most cases involve the choice of a daily activity pattern from a predefined set of available patterns. Models with this structure are usually called tour-base models. Most of the work that has been done recently on activity-based modeling has occurred with these types of econometric models. Well-known models of this type include the Bowman and Ben-Akiva model (2001), the MORPC model (Vovsha et al., 2004a), and the Jakarta model (Yagi and Mohammadian, 2008a and 2008b), among many others, which are all tour-based models, and the Comprehensive Econometric Micro-simulator for Daily Activity-travel Patterns (CEMDAP) model (Bhat et al., 2004), which is not.

Most of the econometric-based models consider daily activity and travel patterns to be composed of a set of tours (Bowman and Ben-Akiva, 2001; Vovsha et al., 2004). The tours are defined as a series of travel and activity episodes that originate and terminate at the home location. The tours are generally grouped into primary and secondary tours; the primary tour includes what is considered the highest priority activity of the day; all other scheduled tours are considered secondary. For an employed individual or student, the primary tour generally contains a work or school episode, whereas the secondary tours generally contain maintenance or leisure activities (Wen and Koppelman, 2000). To construct the tours, most econometric models use the concept of activity priority, based on the purpose of the activity. Work or school activities have the highest priority, followed by maintenance and discretionary activities (Vovsha et al., 2004). In some models, such as the Mid-Ohio Regional Planning Commission (MORPC) model, the priority is also determined by the degree of interpersonal fixity and the duration of the activity, with longer activities and those with other people considered a higher priority.

The tour-based structure is adopted by these models to reduce the choice set within the nested-logit structure. The creation of an activity-travel pattern in which the activities are not grouped into tours is considered a combinatorial problem (Ettema and Timmermans, 1997). To reduce this complexity, pre-defined tour sequences are adopted, such as home-work-home or home-maintenance-work-home, and the models are specified as a nested series of choice models. For example, the model by Wen and Koppelman (2000) first generates the number of household maintenance stops for the day, assigns them to adult household members, and allocates vehicles to each adult household member, with each subsequent choice conditional on the previous one. After this is completed, the tour structure is created by first choosing the number of tours, then for each tour determining if a maintenance stop will be included, and finally choosing from the resulting available tour patterns.

As is evident, models of this type are structured in a highly sequential manner, and the daily activity pattern is chosen at one time. In addition, most of the models use only a limited number of time periods for analysis: for example, morning peak, midday, evening peak, and night, or some other division. This is also done to reduce the number of choices in the model because time of day is an explicit choice. Therefore, models of this type have been criticized for how they capture time of day effects (Ettema and Timmermans, 1997) as well as how they represent the actual scheduling process (Garling et al., 1994) through the sequential structure of the logit model and the all-at-once schedule creation process. Descriptions of some of the main tour-based and nontour-based econometric models follow.

7.4.1 Bowman and Ben-Akiva Model

Bowman and Ben-Akiva (2001) developed a model that considered the daily activity travel pattern as a set of tours and is defined as the travel from home to one or more activity locations and back home again. Tours are subdivided into *primary* and *secondary* tours based on activity priority. Activities are prioritized based on the purpose of the activity. Work activities have the highest priority, followed by work-related, school, and all other purposes. As a rule, within a particular purpose, activities with longer durations are assigned higher priorities. The tour of the day with the highest priority activity (that is, the primary activity) is designated as the primary tour, and others are designated as secondary tours. This modeling framework represents the daily activity travel pattern by three attributes: the primary activity, the primary tour (that is, number, purpose, and sequence of activity stops), and the number and purpose of secondary tours. The tours are represented by the time of day (five time periods), destination (traffic analysis zone [TAZ]), and mode. A prototype of the model was tested in Boston, and it was fully applied in Portland, Oregon (Bowman and Ben-Akiva, 1998).

7.4.2 PB Consult Models

Activity-based models developed for New York, Columbus, and Atlanta (herein called *PB Consult Models*) are based on a person-based mode/destination/time-of-day choice model that is similar to the Bowman and Ben Akiva Model. However, PB Consult Models use more of a *cascading* model approach, first generating mandatory tours, then maintenance tours, then discretionary tours, then intermediate stops on all tours. The residual time window remaining after higher priority tours and activities are generated and scheduled can be used in the generation of subsequent tours and activities (Vovsha et al., 2004).

7.4.3 CEMDAP

The CEMDAP (Bhat et al., 2004) is a microsimulation implementation of an activity-travel modeling system. It is a system of econometric models that represents individual decision-making behavior. The system differs from its predecessors in that it is one of the first to comprehensively simulate the activity-travel patterns of workers as well as nonworkers along a continuous time frame. Given various land-use, sociodemographic, activity system, and transportation level-of-service attributes as input, the system provides as output the complete daily activity travel patterns for each individual in each household of an urban population.

The overall framework adopted in CEMDAP differs from the tour-based models in that it comprises two major components: the generation-allocation model system and the scheduling-model system. The purpose of the former is to identify the decisions of individuals to participate in activities that are motivated by both individual and household needs. The scheduling system uses these decisions as input to model the complete activity travel pattern of individuals. CEMDAP was applied to the synthetic population generated for the Dallas–Fort Worth area.

7.5 RULE-BASED ACTIVITY SCHEDULING MODELS

A modeling framework advanced as an alternative to address some of the behavioral assumptions found in the econometric models is the rule-based or computational process model (Garling et al., 1994). Rule-based models tend to use either simple heuristics to approximate the scheduling process (Roorda et al., 2005) or computational process models that attempt to represent the decision-making process, rather than modeling only the outcomes of the decision-making process as in an econometric model. Much of the theory of the computational process model is based on work by Newell and Simon (1972) in the development of the production system. The production system is a model of cognitive behavior that states, "Individuals' choices are based on their cognition of their environment" (Ettema and Timmermans, 1997). This means that a cognitive process can be represented by a model which contains an individual's memory, including knowledge of the environment and the results of previous interactions with it, rules that operate on that memory, and some currently-known information about the environment. This allows the individual to form resulting thought or to take action, which is then added to memory. The rules are usually introduced as a set of if-then rules, wherein the current information in the system is matched to the conditions of the *if* statement; if matched, the action represented by the *then* statement is undertaken, and the results of the action can be used to update the if-then rules if necessary (Ettema and Timmermans, 1997).

Although less work has focused on research in this area, several models using rule-based frameworks or computational process models (CPMs) have been developed. One of the earliest models developed using the CPM framework, created by Hayes-Roth and Hayes-Roth (1979), was the first to apply the principles to activity scheduling. More recent models include SCHEDULER (Golledge et al., 1994), TASHA (Roorda et al., 2005), and ALBATROSS (Arentze et al., 2000). These models all attempt to specify the process of activity scheduling and are therefore potentially more theoretically satisfying; they are also more policy sensitive, because this type of model is the only one offering policy scenarios that represent actual changes in the scheduling process itself.

7.5.1 SCHEDULER

The SCHEDULER model is an early example of the use of a computational process model in activity scheduling, which updates the earlier Hayes-Roth model to correct certain deficiencies in the decision-making process (Golledge et al., 1994). It attempts to model the learning process by storing the outcomes of each scheduling experience into long-term memory and using those outcomes in the subsequent activity scheduling steps. By this process, the modeled individual can be said to learn about the environment and use the knowledge gained to make more informed decisions as the model progresses. The individual's long-term memory is represented using the concept of the cognitive map which contains the user's imperfect knowledge about the system—activity locations, travel times, and temporal constraints—and is updated through experience (Golledge et al., 1994).

The SCHEDULER model assumes a list of prioritized activities that need to be scheduled, similar to the concept of the project agenda described earlier. The model then schedules these activities according to the given constraints. For this model, the computational process is represented by the attempts to add activities to the schedule in the order of highest priority; if the constraints do not allow the high priority activity to be added, the next-highest priority activity is scheduled (Golledge et al., 1994). As such, the CPM is a simplified version of actual decision-making and conflict resolution behavior as it is based solely on the assumed priorities of the activities and available time in the schedule.

7.5.2 TASHA

A model similar to SCHEDULER is the TASHA model created for travel demand estimation in Toronto, Canada (Roorda et al., 2005). Because the TASHA model is designed as part of a full microsimulation model, activity generation is considered a part of the system, unlike SCHEDULER which operates on previously defined activity agendas. In the TASHA model, schedules are built by generating activities, inserting activities into project agendas, and then inserting the activities from agendas into schedules. The model uses a similar computational process to develop the activity schedule as the SCHEDULER model, because both focus on *a priori* heuristic rules to fit activities into a schedule. In this model, activities are generated according to their observed frequencies in a household travel survey of the area, along with feasible start times and durations generated from observed joint distributions, and added to the individual project agendas. The model uses a large number of logical rules to generate schedules that relate to how and when activities can be added and how scheduling gaps are filled (Roorda et al., 2005).

The TASHA model uses a fixed priority order to insert activities from the agenda to the schedule. Priority in this context relates to the utility of the project: it assumes that work activities are the highest priority, followed by school, then joint other and shopping activities, and, finally, individual other and shopping activities. The activities are added to the schedule in this priority order. In addition, because activities are generated randomly, conflict resolution becomes an important part of the model. Conflict resolution in TASHA is handled by shortening activities, shifting activities, doing both, or splitting the activity. If there is no feasible way to resolve the conflict, the new activity is dropped from the schedule. Similar to the SCHEDULER model, the cognitive process is represented in TASHA by the scheduling heuristics based on the activity priority and whether it fits into the schedule. The TASHA model, however, adds an activity generation component and uses somewhat more realistic scheduling rules, although the rules are still based more on assumptions of how the scheduling process functions than on observations of the process itself.

7.5.3 ALBATROSS

Another currently operational rule-based system is the ALBATROSS model, developed to estimate travel demand in the Netherlands (Arentze and Timmermans, 2000). This model, like the TASHA model, is based on travel survey data. Additionally, it uses a schedule skeleton that represents the routine activities of the individual. These routine activities are not changed or modified, and the attributes of these activities, including location, time of day, and duration, are received as given. Unlike TASHA, however, it uses a complicated series of decision rules to build the complete activity schedule, rather than generating the activities based on their distributions. ALBATROSS represents another attempt to create a computational process model of scheduling behavior.

The model is designed as a microsimulation model at the household level for a one-day scheduling period. It simulates activity scheduling for individuals in random order and, for scheduling within households, uses the other household members' current schedules as input. The model begins by deciding on the mode chosen for travel to work, which conditions all other decisions in the model. Next, the model decides whether or not to add an activity of each pre-defined type to the schedule skeleton, and if added it chooses the duration and party composition. Afterward, a general time of day for each activity is chosen, the activities are linked into tours, modes are assigned to the tours, and locations are chosen for each activity. These decisions are made sequentially, with the previous choices and spatio-temporal constraints conditioning the decision. Like the other models reviewed, ALBATROSS assumes a fixed activity priority with all modeling stages following this priority assumption. For example, the decision to add an activity proceeds from high- to low-priority activities, as do the time-of-day and other attribute choices.

All decisions in ALBATROSS are made using decision trees to represent the choice process. The decision trees meet the conditions for specifying a production system of cognitive behavior (Garling et al., 1994), because they represent a mutually exclusive set of condition-action rules. The decision-making process follows the previously listed sequence, according to what was felt to be the closest representation of decision-making behavior: the mode to work is the most important choice, followed by adding flexible activities and specifying their attributes. The ALBATROSS model represents an important step in developing a complete model of activity scheduling behavior.

7.5.4 Dynamic Scheduling Models

The fundamental concept underlying the framework of dynamic activity scheduling models is to try to account for scheduling dynamics, including activity scheduling and rescheduling, and to accomodate the generation of unexpected events and their impacts on the activity schedule.

Recent dynamic scheduling models include AURORA (Joh et al., 2004) and the Agent-based Dynamic Activity Planning and Travel Scheduling (ADAPTS) model (Auld and Mohammadian, 2009). The ADAPTS framework simulates the dynamic process of schedule formation and accounts for the varying interdependencies and possible differences in planning times between the various attributes. For example, an activity can be added and a location chosen at the same time, but the timing for the activity can be left open. If the timing is decided later, it will depend on the location choice. However, the timing might not depend on the location choice at all, but could be planned first or at the same time. The detailed framework for the ADAPTS activity scheduling model is presented in Figure 7.1.

At the conceptual level, the model can be thought of as splitting the activity scheduling process into three distinct phases. The first phase is activity generation. In this framework, activity generation refers only to the highest-level decision: whether or not to add an activity of a certain type and when to plan the attributes of the activity. The second phase of the model framework is activity planning. In this phase, the actual values of the various activity attributes are specified. This framework allows the attributes to be determined in any order and attributes planned later are dependent on the already-planned attributes. Finally, the last phase of the framework is activity scheduling, in which the activities are added to the planned schedule and conflicts resolved. This framework presents activity scheduling as a dynamic process, completed over time, with the final executed schedule resulting from a series of interdependent planning and scheduling decisions.

7.6 MICROSIMULATION IN ACTIVITY-BASED MODELING

Econometric, rule-based, and dynamic scheduling models are generally implemented within an overall microsimulation framework. This means that each of the decision-making units located within the study area, whether an individual, household, or some other unit, is represented individually. The econometric activity pattern selection model or rule-based activity scheduling framework is then applied to develop each individual's activity patterns. The activity pattern defines the individual's movements from and to each activity location and, therefore, determines the individual's travel pattern throughout the day. The combination of all travel patterns from each household or individual creates the regional travel demand. Whether the microsimulation follows econometric or rule-based activity modeling, generally the same fundamental procedures are followed to create the simulation. Either model starts by initializing the study population, usually through some type of population synthesis sub-model

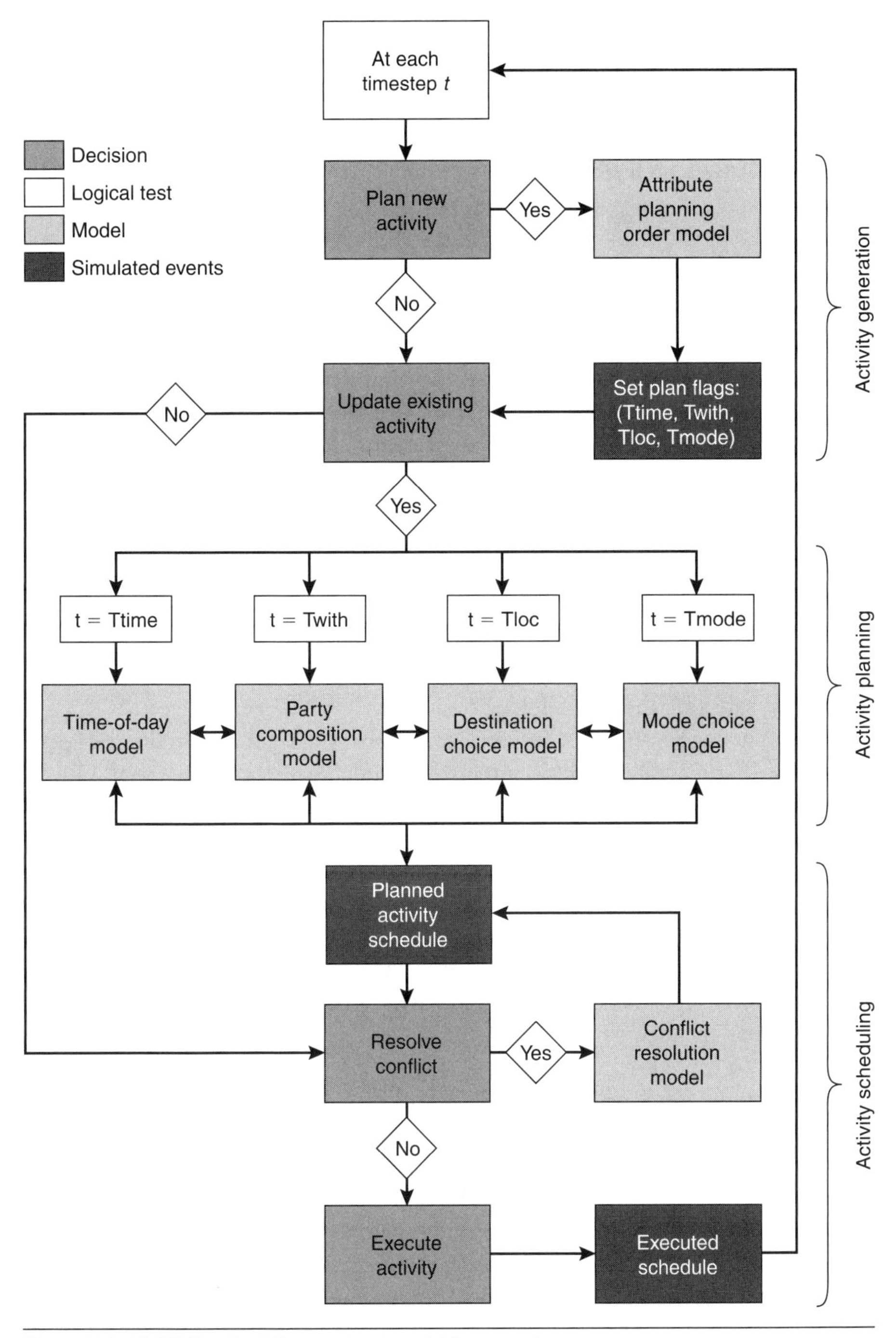

Figure 7.1 ADAPTS scheduling process model framework

as described in work by Beckman et al. (1996) and others. Next, the activities in the individual's travel patterns are generated; the destinations, travel modes, party composition, duration, and start time are planned. Finally, the travel pattern is extracted.

7.6.1 Population Synthesis

Many population synthesizers currently in use for various activity-based microsimulation models are based on the method proposed by Beckman et al. (1996) for use in the TRANSIMS model, which utilizes the statistical technique of iterative proportional fitting. This procedure matches exact large-area multi-dimensional distributions of select variables from the Census Public Use Microdata Sample (PUMS) files with small-area marginal distributions from the Census Summary files to estimate the multidimensional distributions for the small areas (Beckman et al., 1996). The procedure works for any situation whose exact multidimensional distribution and marginal totals are known; it is not limited to using Census data files. A process of this type is necessary for all population synthesizers because confidentiality and data aggregation issues preclude most agencies from releasing multidimensional distributions for the small areas typically needed for most models. Other models, however, can use different data sources or even estimated data, as is the case in the MORPC model (PB Consult, 2003). In general, the population synthesis is typically carried out in two stages: fitting the multidimensional distribution and selecting the households.

The first stage in the synthesizing procedure is developing the multidimensional distribution matrix (Beckman et al., 1996). This stage uses the technique of iterative proportional fitting (IPF), which has long been used in many fields (Deming and Stephan, 1940) and assumes that the correlation structure among the variables is the same for both the larger and the smaller areas of which it is constituted. In the IPF procedure, an initial seed distribution from the disaggregate data is fitted to known marginal totals for an area. The difference between the current total and the marginal total for each category of the current variable is then calculated, and the cells of that category are updated accordingly. This process continues for each variable until the current totals and the known marginal total match to some level of tolerance, producing a distribution that matches the control marginal totals and is as similar to the initial distribution as possible. For a more detailed overview of the IPF procedure see the paper by Beckman et al. (1996) and the technical report by Hobeika (2005).

The second stage in the population synthesis is the creation of actual synthesized individuals or households. It is accomplished in one of several ways, depending on certain other model characteristics. The simplest method is to calculate a selection probability for choosing each household from the actual household microdata file for the large area. The calculation is based on the need for households of that type as determined from the final calculated multidimensional distribution table for the small area and the number of households fitting the demographic type in the household list for the area (Bowman, 2004). The households are then selected randomly according to their selection probabilities for each zone as is done for the TRANSIMS model when the PUMS data is used as input (Beckman et al., 1996).

7.6.2 Activity Generation and Planning

Generating the activities is generally a stand-alone model component within rule-based models (as well as the CEMDAP model), whereas in econometric or tour-based models the activity generation is handled through the selection of the daily activity pattern. Many models that have been implemented either generate activities randomly-based on probability distributions derived from activity survey data as in TASHA (Roorda et al., 2005), or they use some type of choice model, such as in CEMDAP (Bhat et Al., 2004), or a decision tree model such as in ALBATROSS (Arentze and Timmermans, 2000), to determine if an activity from a given set of activity types will be added.

In tour-based models, the choice of the daily activity pattern is often a top-level decision within a nested-logit framework. An example is shown in Figure 7.2 which is taken from the Jakarta model (Yagi and Mohammadian, 2008a).

This figure shows a top-level decision: to stay home or leave home during the day. If it is decided to leave home, a choice is made among different out-of-home activity patterns. Additional levels are generally included below this one to select the attributes of each activity: for example, each activity pattern would include further decisions to add subtours. In this case, the subtours are considered separate options in the same choice level, that is, home-work-home with at least one maintenance activity subtour (HWH − 1 + M) and home-work-home with one discretionary activity subtour (HWH − 1 D) exist on the same level. Each decision in a structure of this type is *conditional* on the lower-level decisions and *conditions* the upper-level decisions through the use of the logsum variable, which represents the utility from the lower-level choices based up to the higher-level decisions. It becomes obvious that fixed systems of this type, which specify the entire daily activity pattern with all attributes at once, quickly come to involve a large number of potential choice outcomes.

Rule-based models, in general, try to simplify this process by using heuristics or decision rules to represent the cognitive limitations of individuals in activity planning. These models represent the fact that it is infeasible to assume that an individual potentially could consider all the choice possibilities discussed. Simplifying rules are therefore used to represent a planning process that incrementally builds the daily activity schedule rather than chooses the optimal fully-formed plan. These rules can be either rule-of-thumb type heuristics, such

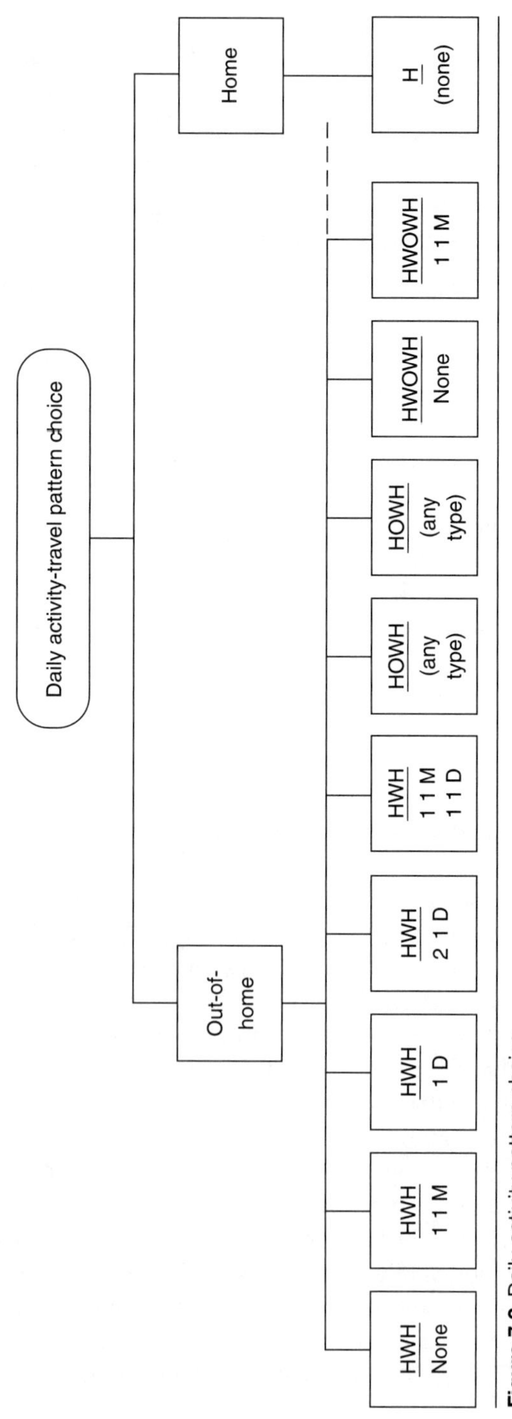

Figure 7.2 Daily activity pattern choice

as in the SCHEDULER model (Garling et al., 1994), or data-derived rules for each planning step, such as those represented by the many decision trees in the ALBATROSS model (Arentze and Timmermans, 2000). Many rule-based models still include econometric or discrete choice models for the selection of individual activity attributes (a nested-logit mode choice model); alternatively, a hazard-based duration model can be used to plan attributes. However, the choice of the overall pattern is represented not as a utility maximizing problem, but rather as the outcome of a process based on limited information that does not always yield an optimal activity-travel pattern.

7.6.3 Activity Travel Execution

The final, and least well-developed, step of activity-based modeling is the schedule execution stage. This is analogous to the traffic assignment step in a traditional four-step model, wherein the travel to scheduled activities is assigned to a transportation network. Frequently, the disaggregate trips and tours achieved through the activity-based scheduling model are simply aggregated to form an O-D matrix of trips to assign to the traffic network. Unfortunately, this process negates many of the advantages of using activity-based analysis. Most importantly, aggregation to an O-D matrix destroys the individualized trip chain information generated by the activity-based model; it also eliminates much of the time-of-day dimension of the travel patterns. Therefore, recent work has focused on ways of integrating the disaggregate output of an activity-based model with traffic assignment, thereby taking advantage of the large amount of individual information that comes out of the activity-based model. Many traffic microsimulation models, most utilizing the dynamic traffic assignment procedure, have been developed in recent years, including TRANSIMS, MATSIM, DYNAMIT, and VISTA. These models can microsimulate detailed traffic assignment, sometimes propagating individual vehicles and adjusting their routes through the transportation network. Models of this type, if combined with an activity-based model, would allow all individual-level information to be retained and utilized. An early effort to accomplish this was the development of TRANSIMS, which combines a detailed traffic simulation with some simplified activity pattern generation methods that replicate observed activity patterns for a study area from survey data. Rieser et al. (2007) have also recently coupled an activity-generation model with the traffic simulation in the MATSIM model.

Overall, activity-based models represent an improvement over the aggregate four-step procedures used in the past. Activity-based models use many new modeling techniques, microsimulation methods, and advances in technology and computing power to enhance the realism of travel demand models. Much has been accomplished since activity-based modeling was first proposed; however, much work remains.

7.7 SUMMARY

7.7.1 Future of Rule-Based Activity Scheduling Models

Activity-based microsimulation models can be improved in many areas. Most of these models were developed only to demonstrate the ability to create daily schedules, but they are not fully specified and consistent behavioral models of travel. Areas needing further research include the representation of scheduling dynamics, utilization of behavior process data in activity generation and conflict resolution, extension to a weeklong scheduling period, and implementation of behavioral and learning concepts in route selection. Furthermore, research in analyzing the transferability of behavior rules and verifying the accuracy of the models needs to be initiated.

The most significant improvement needed within current rule-based models—all activity-based models in general—is a more thorough representation of scheduling dynamics. For the most part, currently implemented activity-based models follow a format in which an activity pattern is planned, either through selection of daily patterns or rule-based schedule creation, and then, as a separate step, the activity is executed. This type of analysis, however, ignores the dynamic nature of the activity planning and execution process, whereby activity schedules are modified and adjusted to fit changing individual contexts and needs throughout the execution of the schedule. This implies that schedule planning and execution should be more closely linked, either through a feedback mechanism or some other representation, than they are currently. Recent studies performed in planning and rescheduling by Mohammadian and Doherty (2006), Ruiz and Timmermans (2006), and Ruiz and Roorda (2008), among others, as well as the creation of the AURORA and ADAPTS models, show that research has begun to focus on this issue.

A related problem, as mentioned previously, is that most of the models operate based only on an assumed priority of activities and whether they can be made to fit into the schedule, along with other ad hoc scheduling rules devised to make a realistic schedule (Roorda et al., 2005). The use of newly-collected and analyzed scheduling process data will allow for a much more realistic modeling of many of the decision facets inherent in a scheduling model. This will be especially useful in the activity-generation and conflict resolution stages of the model.

Current models do not take into account many important decision facets of the activity-generation stage (Roorda and Miller, 2005). For example, the collection of process data has shown that differing types of activities have variable planning horizons. This means that certain types of activities can be routine, spontaneous, or planned in advance, and this distinction can have important effects on how these activities are scheduled. Models of planning horizons for various activities have been developed (Mohammadian and Doherty, 2006) but have yet to be implemented within an activity-based modeling framework.

Additionally, the conflict resolution results in TASHA and SCHEDULER are based on assumed activity priority alone. Research on conflict resolution by Roorda and Miller (2005) showed that this is not the case in a significant number of conflicts. Activity conflict resolution is actually a complex decision-making process that is usually simplified in modeling systems by using priority rules alone. Such methods, however, can lead to oversimplifications of actual behavior and to overestimation of the occurrence of high-priority activities; for example, in TASHA, an activity cannot be removed from the schedule once added (Roorda et al., 2005) and in ALBATROSS, an activity is not modified once added (Arentze et al., 2000). Some work has been done on evaluating characteristics of activity conflicts and identifying resolution strategies (Roorda and Miller, 2005). Additionally, Ruiz and Timmermans (2006) estimated a hazard model for activity conflict resolution, but it was applicable only to activities inserted between pre-planned activities and considers only timing and duration changes to the activities. A model that determines rules for how activity conflicts are resolved in more general situations was developed by Auld et al. (2008), but it only estimates four basic resolution strategies and does not evaluate the extent to which modifications take place as was researched by Ruiz and Timmermans (2006).

Another more basic area in which further research is needed is in extending the models to a full week or longer, rather than merely developing a daily activity pattern alone. All the models mentioned focus on daily patterns and, therefore, potentially could miss important day-to-day scheduling process characteristics. Therefore, the model will be developed to take into account weekly schedule creation, because it is closer to the way people actually think about making their schedules, as shown by the large number of activities that are neither routine nor spontaneous but were pre-planned within the weekly period (Mohammadian and Doherty, 2006).

Another area of rule-based microsimulation modeling frequently discussed and conceptualized but rarely implemented is the use of learning behavior in route selection or network assignment and then incorporating the network assignment results into the activity scheduling stage. Activity-based models often create detailed activity schedules using microsimulation but then aggregate the resulting flows into the O-D matrices used in traditional travel demand models for critical time-of-day periods. These O-D matrices are then assigned to the network in the same manner as in traditional models. This omits some detailed flow data from the activity model and, therefore, the activity models are not truly representative of traveler behavior. The model should incorporate individual's perceived travel time in the scheduling stage and allow the individual to learn appropriate route selection from past results.

Recent advances in developing dynamic activity scheduling models attempt to determine the actual activity scheduling process explicitly, rather than just describing the executed activity patterns. The work in developing these types of scheduling process models is being aided greatly by new sources of data that can capture the actual activity scheduling process of individuals (Doherty and Miller, 2000). Modeling the scheduling process dynamically over long time periods represents one of the more significant goals in the field and is currently being pursued by a number of researchers.

The final aspect missing from currently-operational models is an analysis of the transferability of the model and a rigorous verification of the model results. The transferability issue is one of the major criticisms of the traditional model, therefore, it would be valuable to show that activity scheduling behaviors are, in fact, transferable across populations. Additionally, verification is essential to show that models actually work and are more useful and provide better results than current models.

REFERENCES

Arentze, T. A. and H. J. P. Timmermans. 2000. Albatross: A learning-based transportation oriented simulation system. European Institute of Retailing and Services Studies, Eindhoven.

Auld, J. A., A. Mohammadian, and S. T. Doherty. 2008. Analysis of activity conflict resolution strategies. *Transportation Research Record: Journal of the Transportation Research Board*, 2054, 2008, 10–19.

Auld, J. A. and A. Mohammadian. 2009. Adapts: Agent-Based Dynamic Activity Planning and Travel Scheduling Model—A Framework, Proc. of the 88th Annual Meeting of the Transportation Research Board (DVD). Washington, DC. January 2009. *Transportation Research Record: Journal of the Transportation Research Board*, 2054, 2008, 10–19.

Axhausen, K. 1990. A simultaneous simulation of activity chains and traffic flow. In *Developments in dynamic and activity-based approaches to travel analysis*, P. Jones, ed. Aldershot, UK: Avebury, 206–225.

Bhat, C. R., J. Y. Guo, S. Srinivasan, and A. Sivakumar. 2004. Comprehensive econometric microsimulator for daily activity-travel patterns. *Transportation Research Record: Journal of the Transportation Research Board,* National Research Council, vol. 1894, 57–74.

Beckman, R. J., K. A. Baggerly, and M. D. McKay. 1996. Creating synthetic baseline populations. *Transportation Research A*, 30: 415–429.

Bowman, J. L. 2004. A comparison of population synthesizers used in microsimulation models of activity and travel demand. Working Paper, http://jbowman.net/papers/B04.pdf (accessed August 1, 2007).

Bowman, J. L. and M. E. Ben-Akiva. 1996. Activity-based travel forecasting. Proceedings of the activity-based travel forecasting conference, June 2–5.

——— 2001. Activity-based disaggregate travel demand model system with activity schedules. *Transportation Research A*, 35: 1–28.

Boyce, D. 2002. Is the sequential travel forecasting procedure counterproductive? *ASCE Journal of Urban Planning and Development*, 128: 169–183.

Deming, W. E. and F. F. Stephan. 1940. On a least-squares adjustment of a sampled frequency when the expected marginal totals are known. *Annals of Mathematical Statistics*, 11: 427–444.

Doherty, S. T. and E. J. Miller. 2000. A computerized household activity scheduling survey. *Transportation*, 27: 75–97.

Ettema, D., A. Borgers, and H. Timmermans. 1993. Simulation model of activity scheduling behavior. *Transportation Research Record*, 1413: 1–11.

Ettema D.F. and H. J. P. Timmermans. 1997. *Theories and Models of Activity Patterns. Activity-Based Approaches to Travel Analysis*. Oxford, UK: Elsevier.

Garling, T., M.-P. Kwan, and R. G. Golledge. 1994. Computational-process modeling of household travel activity scheduling. *Transportation Research B*, 25: 355–364.

Golledge, R. G., M.-P. Kwan, and T. Garling. 1994. Computational-process modeling of household travel decisions using a geographical information system. Working Paper, UCTC No. 218, The University of California Transportation Center, University of California–Berkeley.

Hagerstrand, T. 1970. What about people in regional science? Papers of the Regional Science Association, 23: 7–21.

Hayes-Roth, B. and F. Hayes-Roth. 1979. A cognitive model of planning. *Cognitive Science*, 3: 275–310.

Hobeika, A. 2005. TRANSIMS fundamentals: Chapter 3, population synthesizer, technical report, U. S. Department of Transportation, Washington, DC, July 2005, available at http://tmip.fhwa.dot.gov/transims/ transims_fundamentals/ch3.pdf (accessed August 1, 2007).

Joh, C.-H., T. A. Arentze, and H. J. P. Timmermans. 2004. Activity-travel rescheduling decisions: Empirical estimation of the aurora model. *Transportation Research Record*, vol. 1898, 10–18.

Kitamura, R. 1996. Applications of models of activity behavior for activity-based demand forecasting. Proceedings of the Activity-Based Travel Forecasting Conference, June 2–5.

Kitamura, R. and M. Kermanshah. 1983. Sequential model of interdependent activity and destination choices. *Transportation Research Record: Journal of the Transportation Research Board*, 944: 81–89.

Miller, E. J. 2005. *Propositions for modelling household decision-making, in integrated land-use and transportation models: Behavioural Foundations*, M. Lee-Gosselin and S. T. Doherty, ed. 21–60 Oxford, UK: Elsevier.

Mohammadian, A. and S. T. Doherty. 2006. Modeling activity scheduling time horizon: Duration of time between planning and execution of pre-planned activities. *Transportation Research A*, 40: 475–490.

Newell, A. and H. A. Simon. 1972. *Human Problem Solving*. Englewood Cliffs, NJ: Prentice-Hall.

PB Consult. 2003. Task 2: Household and population synthesis procedure, PB consult/Parsons Brinckerhoff, report prepared for the Mid-Ohio Regional Planning Commission as part of the MORPC Model Improvement Project. March 12.

Recker, W. W. and M. G. McNally. 1986. An activity-based modeling framework for transportation policy evaluation. In *Behavioral Research for Transportation Policy*, 31–52. Utrecht, The Netherlands: VNU Science Press.

Rieser, M., K. Nagel, U. Beuck, M. Balmer, and J. Rümenapp. 2007. Agent-oriented coupling of an activity-based demand generation with a multiagent traffic simulation. *Transportation Research Record*, 2021: 10–17.

Rindsfüser, G. and F. Klügl. 2005. *The Scheduling Agent—Using SeSAm to implement a generator of activity programs. Progress in activity-based analysis.* Oxford, UK: Elsevier.

Roorda, M. J., S. T. Doherty, and E. J. Miller. 2005. *Operationalising household activity scheduling models: addressing assumptions and the use of new sources of behavioural data. Integrated Land-Use and Transportation Models: Behavioural Foundations*. Oxford, UK: Elsevier.

Roorda, M. J. and E. J. Miller. 2005. *Strategies for resolving activity scheduling conflicts: An empirical analysis. Progress in Activity-Based Analysis*. Oxford, UK: Elsevier.

Ruiz, T. and M. J. Roorda. 2008. Analysis of planning decisions during the activity-scheduling process. Proceedings of the 87th Annual Meeting of the Transportation Research Board, January 21–25, 2007.

Ruiz, T. and H. Timmermans. 2006. Changing the timing of activities in resolving scheduling conflicts. *Transportation*, 33: 429–445.

Shiftan, Y., M. Ben-Akiva, K. Proussaloglou, G. de Jong, Y. Popuri, K. Kasturirangen, and S. Bekhor. 2003. Activity-based modeling as a tool for better understanding of travel behaviour. Paper presented at the 10th International Conference on Travel Behaviour Research, Lucerne, Switzerland.

Van der Hoorn, T. 1983. Development of an activity model using a one-week activity-diary data base. In *Recent Advances in Travel Demand Analysis*, P. Jones, ed. Gower, UK: Aldershot. 335–349.

Vovsha, P., M. Bradley, and J. L. Bowman. 2004. Activity-based travel forecasting models in the United States: Progress since 1995 and prospects for future. Paper presented at the EIRASS Conference on Progress in Activity-Based Analysis, Maastricht, the Netherlands.

Vovsha, P., E. Petersen, and R. Donnelly. 2004a. Impact of intra-household interactions on individual daily activity-travel patterns. Paper presented at the 83rd Annual Meeting of the Transportation Research Board, Washington, DC.

——— 2004b. Model for allocation of maintenance activities to household members. *Transportation Research Record: Journal of the Transportation Research Board*, 1831, National Research Council, 1–10.

Wen, C.-H. and F. Koppelman. 2000. A conceptual and methodological framework for the generation of activity-travel patterns. *Transportation*, 27: 5–23.

Yagi, S. and A. Mohammadian. 2008a. Modeling daily activity-travel tour patterns incorporating activity scheduling-decision rules, *Transportation Research Record: Journal of the Transportation Research Board*, No. 2076, TRB, National Research Council, 123–131.

——— 2008b. Joint Models of Home-Based Tour Mode and Destination Choices: Applications to a Developing Country. *Transportation Research Record: Journal of the Transportation Research Board*, No. 2076, TRB, National Research Council, 29–40.

8

MAXIMUM SIMULATED LIKELIHOOD ESTIMATION WITH SPATIALLY CORRELATED OBSERVATIONS: A COMPARISON OF SIMULATION TECHNIQUES

Xiaokun Wang
Department of Civil and Environmental Engineering
Bucknell University

Kara M. Kockelman
Department of Civil, Architectural, and Environmental Engineering
The University of Texas–Austin

8.1 INTRODUCTION

Econometric models are a powerful tool for analyzing regional issues. Complex models are normally intractable and require special estimation methods. Maximum simulated likelihood estimation (MSLE) techniques have become popular in recent years and are being included in new software releases (such as STATA and

Limdep). It is important that analysts understand the relative performance of simulation techniques under various data circumstances. This is especially true in regional studies because observations are often spatially correlated.

This chapter studies the performance of several simulation techniques with spatially correlated observations. Quasi-Monte Carlo (QMC) methods are found to impose a strong periodic correlation pattern across observations. While some forms of sequencing, such as scrambled Halton, Sobol, and Faure can sever correlations across dimensions of error-term integration, they cannot remove the correlation that exists across observations. When a data set's true correlation patterns clearly differ from the simulated patterns, model estimation can become inefficient; and, with finite samples, statistical identification of parameters can suffer. Within the mixed logit framework, at least, we find that even though observations are correlated, QMCs and hybrid methods are typically preferred to pseudo-Monte Carlo (PMC) methods, thanks to better coverage. These findings offer an important supplement to existing studies of spatial model estimation and should prove valuable for future work that requires simulated likelihoods with spatially correlated observations.

Simulation is an indispensable tool in many research fields for studies of complex systems. Thanks to enhanced computational power, simulation has become more popular in recent years. One important application is the estimation of statistical models in which analytical derivation is impractical (for example, when a likelihood function involves a multidimensional integral). In regional studies, a typical application is the estimation of complex spatial econometric models, especially those that are nonlinear in nature. In this study, simulation techniques are explored within the framework of a mixed (random-coefficients) logit model in which observations are spatially correlated.

McFadden (1989) first applied MSLE for discrete choice. Ideally, simulated draws should be randomly chosen points from the given (typically standard uniform) distribution. In practice, however, random numbers have to be generated using a deterministic routine. For example, the widely-used PMC method generates numbers in a deterministic fashion, but in a way that appears to be random under simple statistical tests (Niederreiter, 1992). This type of simulation causes integration errors[1] to converge on zero at a rate of $\sqrt{N}$, wherein N is the number of observations (Spanier and Maize, 1994).

In contrast to PMC, QMC methods do not generate apparently random values. Their major advantage is that they are uniformly distributed with better coverage, offering a faster convergence rate for the upper bound of integration error (Bhat, 2003). There are several ways to generate QMCs. Commonly used approaches are the Faure sequence (Faure, 1992), Sobol sequence (Sobol, 1967), Halton sequence (Halton, 1960), and Halton's various scrambled and shuffled versions (Braaten and Weller, 1979; Hellekalek, 1984; and Hess and Polak, 2003). As Bhat (2003) explains, a scrambled Halton works better than a standard Halton for high dimension integration. One limitation of QMC is that there is "no practical way of statistically estimating the integration error" (Bhat, 2003). Another potential issue, which is discussed in this chapter, is the periodic correlation pattern inherent in QMC. For samples with independent observations, the effect of this correlation pattern is not significant; however, it can become a problem when the model involves correlation across observations.

Bhat (2003) suggests an attractive alternative that combines PMC and QMC—randomized QMC. Bhat (2003) proposed shifting a Halton sequence by adding a random number to the standard Halton sequence. In this way, the integration error can be statistically estimated. However, shifting does not break the cycling of a standard Halton sequence. Thus, it does not help remove the correlation pattern (across observations) in QMC. Hess and Polak (2003) suggested another hybrid method—a shuffled Halton sequence. By randomly shuffling (reordering) a short Halton sequence for each observation, sequences avoid correlation in higher dimensions while speeding sequence generation time. However, because the shuffled Halton sequence uses the same numbers (in different orders) for all observations, the simulation noise is likely to increase with sample size instead of canceling. This method can lead to inconsistent estimates as already mentioned.

In summary, many studies seek improved simulation techniques to produce unbiased, consistent, and maximally efficient estimates. Most existing evaluations of simulation techniques focus on asymptotic properties using assumptions of interobservational independence (Ben-Akiva et al., 2001; Bhat, 2003; and Hess and Polak, 2003). When there are repetitive choice experiments, researchers tend to add an individual effect into the model specification (Hensher and Green, 2003).

In regional studies, observations are often spatially correlated. For example, in models of land use change, land development of a specific site often depends on neighborhood conditions. In other data sets (for example, traffic counts), observations are interrelated through network connectivity. Spatial and serial autocorrelation in such situations are systematic. In these cases, simulation techniques relying on an independence assumption can be problematic; simply specifying unit-specific random effects is not adequate either.

Thus, the objective of this study is to explore various simulation techniques in circumstances wherein continuous and systematic correlation across observations exists. Six simulation methods are compared: PMC, standard Halton, scrambled Halton, shuffled Halton, and two other hybrid methods referred to as *long shuffled Halton* method and *randomized shuffled Halton* sequence. Two other QMC sequences, Sobol and Faure, are discussed briefly also, but not compared in detail because of their similarity to standard Halton sequences.

The long shuffled Halton sequence has a generation rule similar to the shuffled Halton. Instead of shuffling the same sequence for all observations, the long shuffled Halton approach shuffles a long standard Halton sequence and then splits it for each observation. The randomized shuffled Halton is a hybrid version of Bhat's (2003) shifted Halton sequence and Hess and Polak's (2003) shuffled Halton sequence. It adds a uniformly distributed random number to each value in a shuffled sequence. The relative performance of these six techniques is then compared based on theory and an empirical study using synthetic data. As this chapter highlights, standard and scrambled Halton methods import periodic correlations across observations. Thus, in situations in which correlated observations exist, Halton and scrambled Halton sequences can cloud or counteract the true correlation pattern, leading to inconsistent (or at least inefficient) estimates.

The following sections describe existing studies of mixed logit model estimation and simulation techniques, along with potential problems for different simulation methods. These methods are applied to a synthetic dataset exhibiting spatial correlation. In addition, the results of simulation methods are compared.

8.2 MIXED LOGIT MODEL

The mixed logit model, also called the random parameters or kernel logit model, has been widely used for a number of years. It has been applied to topics in finance, biometrics, and social science, among others. Its application in regional science is broad. For example, Rouwendal and Meijer (2001) used it to study preferences for housing and jobs, Wang and Kockelman (2006) studied the spatial and temporal evolution of land cover in urban areas, and Hensher and Greene analyzed urban commuting.

Initially, the mixed logit model was designed to incorporate heterogeneity and correlation across alternatives. Thanks to its great flexibility, its applications were extended. Srinivasan and Mahmassani (2005) proved that mixed logit models are capable of approximating any random utility model. And Greene (2002) notes how mixed logit models can conveniently incorporate panel effects.

McFadden and Train's (2000) work depicts a general formulation for mixed logit models, in which parameters are assumed to be randomly distributed with a density distribution function $f(\beta|\theta)$. β denotes the random coefficients, and θ parameters characterize their random distribution. For observation i, the probability of choosing alternative q is:

$$P_{iq} = \int \left(\exp(x_{iq}'\beta) / \sum_{p} \exp(x_{ip}'\beta) \right) f(\beta|\theta) d\theta \tag{8.1}$$

where x_{iq} is the vector of explanatory variables. This integration can be approximated using simulation:

$$SP_{iq} = \frac{1}{R} \sum_{r=1}^{R} P_{iq}(\beta^{ir}) \tag{8.2}$$

where SP_{iq} is the simulated probability for observation i's choice of alternative q, β^{ir} is the rth of R draws for observation i from its density $f(\beta|\theta)$. By construction, SP_{iq} is an unbiased estimate of P_{iq}, and, therefore, the simulated likelihood value SL_i is also an unbiased estimator of the true likelihood value L_i: $E(SL_i) = L_i$.

The log likelihood of observation i choosing alternative q is:

$$\ln L_i = \sum_{p=1}^{J} d_{ip} \ln P_{ip} \tag{8.3}$$

where d_{ip} equals 1 if observation i chooses alternative q and 0 otherwise. Unfortunately, the log transformation leads to bias in the simulated log likelihood. As Train (2003) shows:

$$\ln SL_i = \ln L_i + (E(\ln SL_i) - \ln L_i) + (\ln SL_i - E(\ln SL_i)) \tag{8.4}$$

The second part of this expression ($E(\ln SL_i) - \ln L_i$) indicates the bias caused by the transformation, denoted here as B. A second-order Taylor expansion suggests that:

$$B = E(\ln SL_i) - \ln L_i \approx -\frac{1}{2L_i^2}\mathrm{var}(SL_i) \tag{8.5}$$

As Train (2003) explains, var (SL_i) is inversely proportional to the number of draws, R. Therefore, when R is fixed, this bias B accumulates with sample size, making MSLE inconsistent. Only with an increase in R does bias disappear.

The third part of Equation (8.4), $\ln SL_i - E(\ln SL_i)$, defines the simulation noise which is caused by the difference between draws actually used for SL_i and its expectation over all possible draws. It is referred to here as C, which also has a variance inversely proportional to R. As long as SL_i is an unbiased estimator of L_i, the major concern is to reduce the summation of B and C over the entire sample. With PMC, generally, this can be achieved by increasing R. The remaining issue is to find the simulation technique that achieves these goals most efficiently. Train (2003) suggests that if draws are negatively correlated within one observation and/or across observations, the sample summation of B and C tends to fall. Train (2003) supported this assertion by using R = 2 as an example.

The variance of:

$\hat{t} = (t(\varepsilon_1) + t(\varepsilon_2))/2$ is $(var(t(\varepsilon_1)) + \mathrm{var}(t(\varepsilon_2)) + 2\,\mathrm{cov}\,(t(\varepsilon_1)), t(\varepsilon_2))/4$.

A negative $\mathrm{cov}(t(\varepsilon_1), t(\varepsilon_2))$ reduces the variance of average. Negative correlation within a given observation's draws, which can reduce B, actually provides good coverage. Correlation across observations, which influences the total variance of simulation noise C, depends on the relationship across draws used for different observations. Based on all these relationships, studies of simulation techniques seek a simulation method that provides good coverage and negative correlation across observations.

8.3 SIMULATION TECHNIQUES

8.3.1 Standard Halton Sequence

As previously mentioned, Halton sequences (Halton, 1960) are deterministic series of values between zero and one that provide uniform coverage in this number space. Each sequence is generated via a prime number, also called its base. Bhat (2003) explains the generation of Halton sequences with formulations and Train (2003) illustrated how a standard Halton sequence draws numbers cyclically on the [0,1] interval. To make the following explanation clearer, Train's (2003) illustration is briefly described as:

> *Using prime number 3 as an example, the first cycle divides the interval into three sections, their left edges labeled A, B, and C (see Figure 8.1). Then each section is divided into three parts, and each subsection is labeled lexicographically as for the first cycle. The Halton sequence is generated with a modified alphabetical order: B for 1/3, C for 2/3, BA for 1/9, BB for 4/9, BC for 7/9, CA for 2/9, CB for 5/9, and CC for 8/9. The draws continue cycling in this way, filling in the intervals' gaps. Zero and duplicated points (for example, AA = A = 1, AB = B, AC = C) are ignored.*

Train (2003) noted that Halton sequences exhibit a negative correlation within one observation's sequence as well as across adjacent observations. Thus, Halton sequences can perform much better (in reducing simulation bias and noise) than independent draws. However, one potential issue is the incomplete consideration of cross-observation correlation. When additional correlation terms are considered, the conclusions can change. For example, if $R = 3$, cov $(t(\varepsilon_1), t(\varepsilon_2)) < 0$ and cov $(t(\varepsilon_2), t(\varepsilon_3)) < 0$, it is quite possible that cov $(t(\varepsilon_1), t(\varepsilon_3)) > 0$. If this positive value exceeds the sum of those two negative values, the final variance will be larger than the variance resulting from independent draws. Unfortunately, this is quite common with Halton sequences. As illustrated in Figure 8.1, a Halton sequence cyclical draws pick up the spaces left by prior draws and ultimately pick positions close to a series of earlier draws. Thus, it is reasonable to expect positive periodic correlations. Wang and Kockelman (2006) noted this pattern across draws and touched on it in their mixed logit models of land-use change based on satellite imagery. In their study, the correlation across observations is discussed based on the correlation between generated sequences. Here, the correlation across observations' simulated probabilities is examined in depth using a method Train (1999) described: A single observation is treated as 1000 different observations, so that

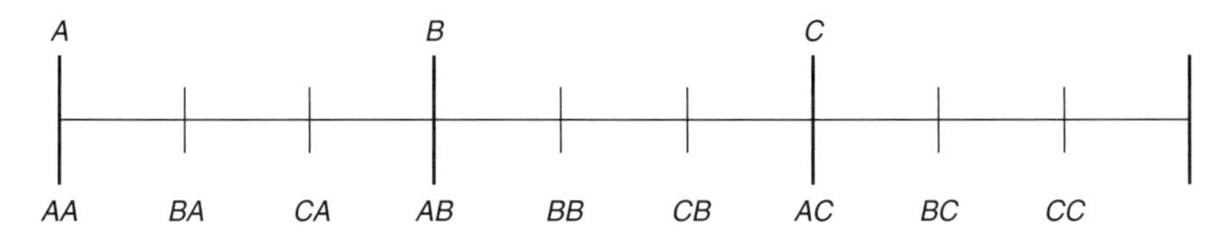

Figure 8.1 Segments for standard and scrambled Halton sequences (from Train (2002)).

the variation pattern in variables is avoided. In this way, the correlation across observations' simulated probability is exclusively generated by the simulation technique.

The difference between this method and the one described by Wang and Kockelman (2006) is that when calculating correlations between two sequences of random numbers, the order of the numbers is important. For example, consider a shuffled Halton sequence in which the randomly-ordered sequences show no correlation across observations. Simulated probabilities, in contrast, diminish the re-ordering effect by averaging a transformation of the simulated values over the set of draws:

$$SP_{iq} = \frac{1}{R}\sum_{r=1}^{R} P_{iq}(\beta^{ir})$$

This means that when using simulated probabilities, the order of r has no influence. From this perspective, Train's (1999) method indicates that shuffled Halton sequences result in perfectly correlated simulated probabilities.

To illustrate the effect of simulation technique on observation correlation, a simple model specification was used: a three-alternative mixed logit model with two alternative-specific constant (ASC) terms and one alternative-specific variable. The alternative-specific variable was generated from a standard uniform distribution and has the fixed coefficient 1. The first alternative serves as the base or reference alternative and, thus, has a fixed ASC of zero. Alternative two's ASC is fixed as 0.5. The ASC for alternative three is normally distributed with mean 0.2 and standard deviation 2.

Figure 8.2 shows the correlation pattern generated using different numbers of draws (R) from the same base of 3 for alternative 3. The vertical axis shows correlation of simulated probabilities, $corr(SP_{\bullet,q}, SP_{\bullet+k,q})$, where $SP_{\bullet,q}$ is the vector of all observations' simulated probabilities for choice of alternative q. (Figure 8.2 shows the pattern for $q = 3$; the $q = 1$ and 2 patterns are virtually the same.) The numbers on the horizontal axis indicate the distance between observations, for example, k. That is, 1 indicates adjacent observations, whereas 3 means every third observation. Similar to Train's (1999) findings, the correlation between adjacent observations is approximately -0.4. However, correlation between pairings

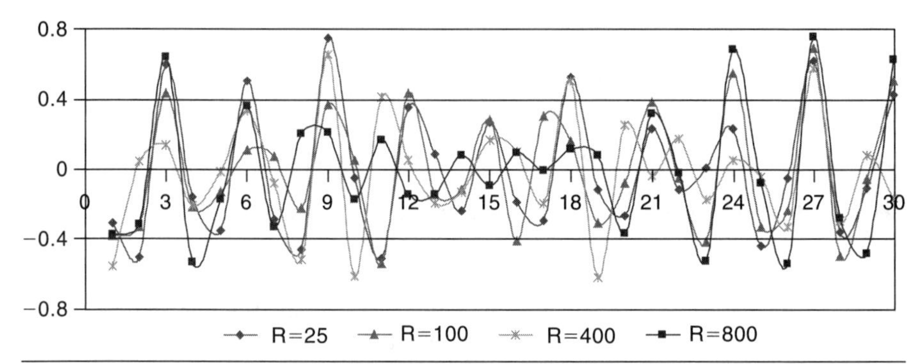

Figure 8.2 Observation correlation generated by Halton sequence with different numbers of draws (base = 3)

Note: $Y = corr(SP_{\bullet}, SP_{\bullet+k})$, $X = k$

of every third observation jumps to +0.6. In fact, with $R = 25$, the same results emerge for all numbers that are multiples of 3. As the number of draws increases, the variations in correlation values soften a bit, but for certain numbers (for example, $k = 27$), the correlation remains quite high.

The base 3 results are similar to those seen in Halton sequences generated using other prime numbers (for example, other bases). As shown in Figure 8.3, the correlations across observations always present approximate periodic patterns with multiples of the bases always associated with high positive values, and the peak correlations can be high. For example, when the base is 5 and R equals 100, the correlation between observations and their 25th nearest observation is 0.975. This means that Halton sequences impose a strong correlation pattern on observations' simulated probabilities. Within a given model specification, the cycle and the magnitude of this periodic correlation depend on its base and the number of draws.

When observations are independent, summation can cancel out these negative and positive correlations, making this effect of correlation across observations insignificant. Meanwhile, the good coverage of a Halton sequence within an observation helps to reduce the value of Equation (8.5). Thus, when observations are independent, the Halton sequence is generally preferred to PMC. Munizaga and Alvarez-Daziano (2001), Bhat (2003), and Hess and Polak (2003) all confirm this conclusion.

However, if observations are correlated, this correlation pattern imposed by use of Halton sequences becomes more problematic.

8.3.2 Scrambled Halton Sequence

Bhat (2003) proposed application of scrambled Halton sequences to estimate mixed logit models. There are many ways to scramble a Halton sequence (Braaten and Weller, 1979; Hellekalek, 1984). The scrambling method discussed is the

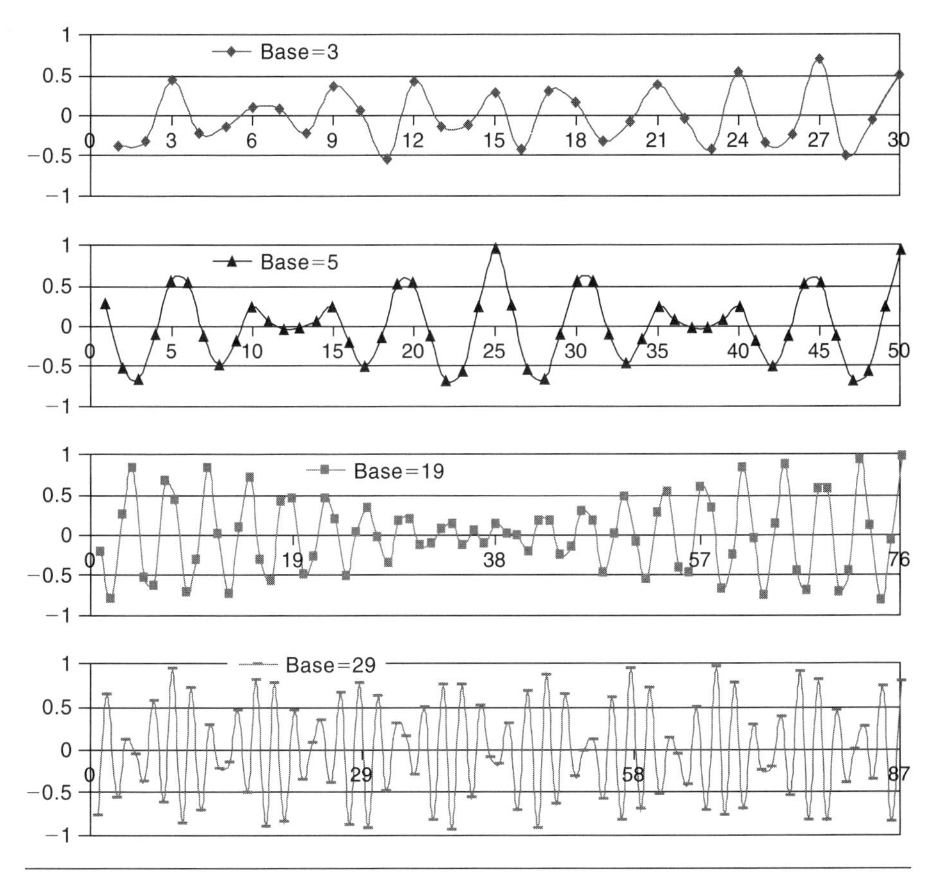

Figure 8.3 Observation correlation generated by Halton sequences with different bases (R = 100).
Note: Y = $corr(SP_{\bullet}, SP_{\bullet+k})$, **X** = k

one proposed by Braaten and Weller and applied by Bhat and Train (2003). By re-ordering coefficients in the number generation rule, scrambling actually exchanges numbers' positions in the original standard Halton sequence (Bhat, 2003). As can be derived from the rule, the first several numbers in the sequence only exchange positions in their near neighborhood. As the sequence length increases, the exchanged positions are farther apart, but then the values of those numbers do not differ significantly. In other words, although scrambling can disrupt the correlation across dimensions (for example, sequences generated with different prime numbers) within each scrambled Halton sequence, an unbroken correlation pattern remains. Compared to the original Halton sequence, a scrambled Halton sequence simply cycles in a different way.

Train's (2003) illustration helps the drawing rule of a scrambled Halton sequence. Using Figure 8.1 further, a scrambled sequence is obtained by reversing the order of B and C. That is, the listing becomes C, B, CA, CC, CB, BA, BC, and BB. The positions of the first eight numbers in standard and scrambled Halton sequences can be shown as follows:

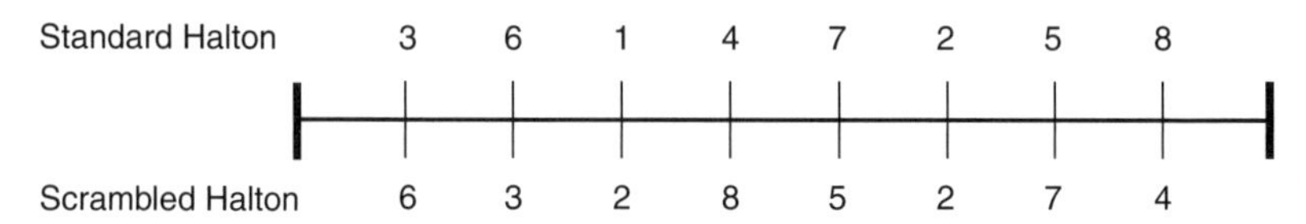

As can be inferred from this analysis, scrambled Halton sequences change number positions only slightly. When considering correlation across observations' simulated probabilities, wherein the ordering of numbers for a single observation does not have any influence, standard and scrambled Halton results differ minimally. In fact, with base 3, the correlation pattern generated by these two methods nearly overlaps. Only with higher prime numbers does scrambling change sufficiently such that the difference between standard and scrambled sequences can be observed. Figure 8.4 shows the correlation across observations induced by a scrambled Halton sequence with base 7. As shown, the sequence still generates a periodic correlation pattern similar to that emerging from a standard Halton sequence. In general, the peaks are slightly lower, but the maximum correlation (for example, at $k = 7$ and its multiples) remains quite high. Therefore, although scrambling removes a standard Halton's correlation across high dimensions, it still has the same problem as a standard Halton sequence when it comes to interobservation correlations.

The main features of a scrambled Halton sequence are representative of several other QMCs, including Sobol and Faure sequences. These two sequences remove

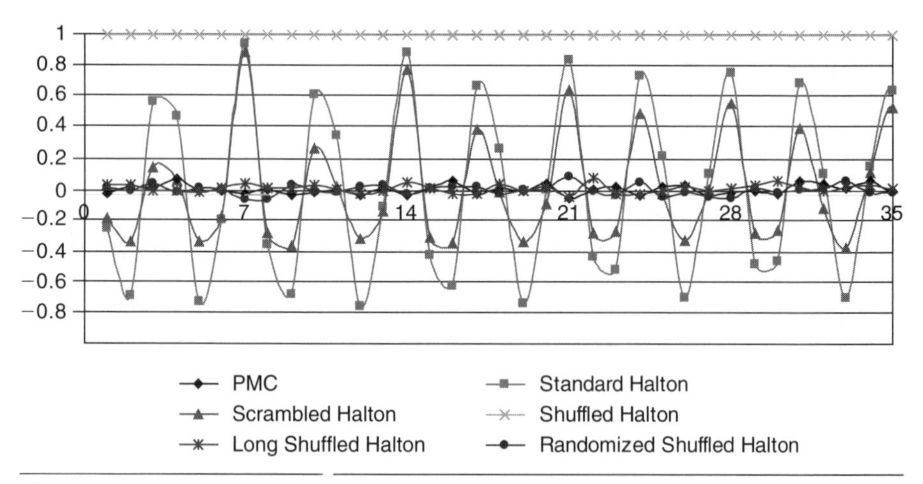

Figure 8.4 Observation correlation generated using different simulation techniques (base = 7, $R = 100$).

Note: $Y = corr(SP_{\bullet}, SP_{\bullet+k})$, $X = k$

correlation across high dimensions in a slightly different way (though within each dimension they work essentially the same as a Halton sequence). In other words, from the perspective of correlation, Sobol and Faure sequences function similar to scrambled Halton sequences. Thus, the scrambled Halton sequence is used here as an example of this type of QMC simulation.

8.3.3 Shuffled Halton Sequence

Hess and Polak (2003) first proposed the use of the shuffled Halton sequence[2] in a mixed logit model. Compared to a scrambled sequence, the generation of a shuffled sequence is more straightforward. The same Halton sequence is randomly shuffled for different observations and alternatives. That is, if there are Q alternatives, and each has just one random coefficient and N observations, a total of $Q \times N$ short sequences need to be generated. If each sequence contains R draws, with a standard or scrambled Halton sequence, Q long sequences, each having $N \times R$ numbers, need to be generated. This means that with large N and R, these two approaches require considerable memory and computation time. With a shuffled Halton sequence, only one standard Halton sequence of length R needs to be generated. This short sequence is shuffled $Q \times N$ times.

One simplified way to shuffle a sequence of numbers is to generate a vector of uniformly distributed random values and sort the sequence according to their order. There are $R!$ ways to permute a sequence containing R different numbers, and in most applications R is at least 50. Therefore, normally $Q \times N$ is much less than $R!$, making these $Q \times N$ sequences uncorrelated (Wang and Kockelman, 2006). This shuffling process can effectively disrupt correlations between alternatives. However, since this method uses the same sequence of numbers, the asymptotic properties can be problematic. In the perspective of correlation across simulated probabilities, a shuffled Halton sequence imposes *perfect* positive correlation across observations, as shown in Figure 8.4. This implies that shuffling a single Halton sequence for all observations is likely to cause a higher estimator variance through simulation noise. It should be less efficient than standard and scrambled Halton sequences (but not necessarily less efficient than PMC, thanks to its better coverage). When shuffling is used for correlated observations, this strong correlation pattern can obscure the true correlation patterns and harm prediction.

Therefore, shuffled Halton sequences should be used with some caution. The initial position and sequence length must be chosen carefully to ensure that the sequence has optimal coverage and does not obscure any behavioral relationships of interest.

8.3.4 Long Shuffled Halton Sequence

To provide a more comprehensive comparison, two new methods are proposed and used here. The first one is the long shuffled Halton sequence; so called because when compared to Hess and Polak's shuffled sequence, the original Halton sequence has a length $N \times R$. The entire long sequence is first shuffled, then divided into N segments for those N observations instead of first dividing and then shuffling. The reason for this is, as discussed previously, reordering a sequence within an observation does not influence the simulated probabilities. This means if a sequence is first cut and then shuffled, in terms of observation correlation, it is equal to using the original sequence and the correlation between simulated probabilities is still not broken. Therefore, its coverage of each observation's distribution is not as uniform as that of a standard or scrambled Halton sequence. However, correlation across observations is low, as shown in Figure 8.4. By construction, this long shuffled Halton sequence should perform like PMC. The second method is a randomized shuffled sequence.

8.3.5 Randomized (Shifted) Shuffled Halton Sequence

In the second method, a shuffled Halton sequence is randomized through shifting. A random number (drawn from a standard uniform distribution) is added to each shuffled Halton sequence. (If the resulting value exceeds 1.0, a value of 1 is subtracted.) As Bhat (2003) and Train (2003) describe, this operation preserves the uniform distribution of the sequences in the number space.

As noted earlier, shuffling offers good coverage within each observation. Its problem is the repeated use of the same sequence of numbers. By adding different random numbers to different sequences (though the same random number is used within one sequence), the problems of shuffling are avoided. As shown in Figure 8.4, the observation correlation generated by this method is close to zero. Meanwhile, this method maintains good coverage. With all these characteristics, this method is expected to out-perform PMC and all QMC methods, especially when observations are correlated. And this is, indeed, the case shown here. In the following sections, all five methods of number generation, together with PMC, are applied to a synthetic data set for MSLE.

8.4 CORRELATED SYNTHETIC DATA

A synthetic data set can be developed for any context as long as the data-generating process is not internally inconsistent. To make the correlation pattern more meaningful, the synthetic data is interpreted as a case of land development. In this example, each observation stands for a grid cell of land at a specific location. There are 1500 observations that compose a rectangular area (Figure 8.5). The value of the dependent variable y can be interpreted as the land use type. There are three alternatives: 1 indicates residential use, 2 indicates commercial use, and 3 indicates industrial use.

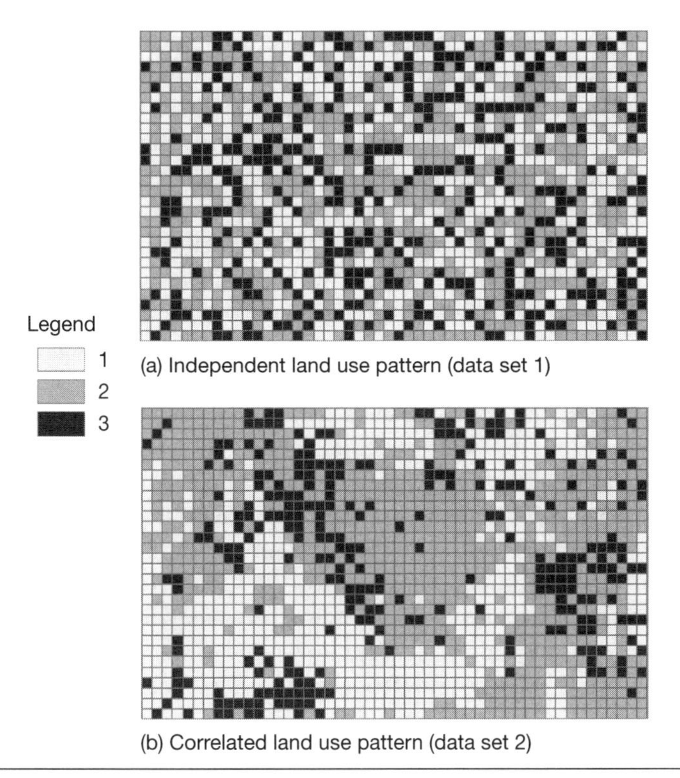

Figure 8.5 Outcomes of the synthetic data

Observation *i*'s random utility for alternative *q* is specified as follows:

$$U_{iq} = \beta_{COST} \cdot COST_{iq} + \beta_{REV} \cdot REVENUE_{iq} + (\alpha_q + \sigma_q \eta_{iq}) + \varepsilon_{iq} \tag{8.6}$$

$COST_{iq}$ is generated from a uniform distribution with mean 1 and standard deviation 0.577. In a land development example, $COST_{iq}$ can be interpreted as the general cost of developing a grid cell and is affected by zoning regulations and construction costs. β_{COST} is a fixed coefficient set to -3. $REVENUE_{iq}$ is drawn from a uniform distribution (mean 1.5 and standard deviation 0.866). It can be interpreted as a property's revenues as determined by floor space prices and/or rents. Its coefficient β_{REV} is fixed to equal 2.

There is also a variable ASC component to the random utility function that is normally distributed with mean α_q and standard deviation σ_q. This component can be interpreted as a constant term plus one normally distributed unobserved error term. Alternative 1's constant term α_1 is -1, and its standard deviation σ_1 is 10. Alternative 2 has $\alpha_2 = 1$ and $\sigma_2 = 5$. Alternative 3 is the base alternative and, thus, has corresponding parameters of 0 and 0. This example considers only correlation across observations: η_{iq} is uncorrelated across alternatives, but within one alternative, it follows a multivariate normal distribution with a mean vector of zeros and unit variance terms. In the first of the two test data sets, there is no correlation across these terms. In the second data set, the covariance terms cov(η_{iq}, η_{iq}) are specified to depend on Euclidean distances between observations i and j, as described later. This is the genesis of the spatial nature of data set 2.

Common in kernel logit models, ε_{iq} follows a standard Gumbel distribution with location parameter equal to 0 and scale parameter equal to 1. It is uncorrelated with the two explanatory variables and η. The total random term per observation and alternative is therefore $\sigma_q \eta_{iq} + \varepsilon_{iq}$. The covariance structure of η_{iq} produces spatial correlation in the random utilities for land use development. Thus, together there are six parameters requiring estimation: β_{COST}, β_{REV}, α_1, σ_1, α_2, and σ_2.

For each observation, the alternative with maximum utility becomes the outcome. Figure 8.5a shows the land use outcomes when the η_{iq} terms are independent across observations (data set 1). For data set 2 (Figure 8.5b), a predetermined correlation pattern is assumed to exist across $\eta_{\cdot p}$ values: The correlation between two observations is calculated as exp($-$ dis_{ij}/10), where dis_{ij} is the Euclidean distance between the centroids of observations i and j, standardized by the length of one unit's edge. Thus, the correlation matrix is of size 1500 × 1500, and the correlation of η values between immediate neighbors is approximately +0.9. The correlation between observations i and j is:

$$corr(U_{iq}, U_{jq}) = \frac{\sigma_q^2 \operatorname{cov}(\eta_{iq}, \eta_{jq})}{\sigma_q^2 + \pi^2 / 6} \tag{8.7}$$

This means that for alternative 1, the correlation between adjacent observations (grid cells) is approximately +0.89. For alternative 2, it is approximately +0.85. (These differ slightly because of differing variance assumptions on their error components.) The correlation here seeks to capture correlations in unobserved attitudes and attributes related to developing parcels within neighborhoods.

8.5 RESULTS ANALYSIS

The five simulation methods discussed, along with PMC, were applied to the synthetic data. Each technique is used seven times with the number of draws R varying from 5 to 50, 100, 200, 500, 1000, and 2000. To avoid a potential identification problem (Ben-Akiva et al., 2001), the fixed coefficient β_{COST} was constrained to equal its true value whereas all other parameter estimates were allowed to change during the MSLE process. All starting values were set to their true values so that problems of local extreme are avoided. For QMCs and their hybrid methods, the bases used to generate $\eta_{i,1}$ and $\eta_{i,2}$ were 3 and 5, respectively.

It can be inferred from this example that 1500 observations is not a large sample. N systematically correlated observations lead to a correlation matrix of size $N \times N$. Thus, a larger sample size will make the synthetic data generation practically unfeasible because of its excessive computational burden. Thus, data sampling variance and bias can contribute to the differences observed between estimates and the true values of the data sets. To account for this, the study used 10 samples (20 data sets total), each drawn from the same distribution described (with 10 exhibiting spatial independence and 10 exhibiting spatial correlation in random error components). Average values of estimates from these samples were compared to their true values. Since each model requires 5 parameter estimates to illustrate the overall accuracy, the root of the sum of squared errors (RSE) of these five parameters were calculated. This measure can be interpreted as the Euclidean distance between the vector of estimated values and the vector of true values: $\sqrt{\|\widehat{\beta} - \beta\|^2}$. From this perspective, the measure serves as a generalized bias.

Table 8.1 shows this generalized estimate of bias with independent observations (for example, data set 1). As expected, standard and scrambled Halton sequences out-perform PMC. With $R = 50$, the simulation bias is already quite low using either method, whereas PMC does not achieve equivalently low bias until R exceeds 200. Also as expected, the shuffled Halton sequence performs better than PMC but not as well as standard and scrambled Halton sequences. The long shuffled Halton sequence performs much like PMC in terms of trend, maximum and minimum biases. The randomized shuffled Halton does not produce good estimates with low values (for example, $R = 50$); however, once R exceeds

Table 8.1 Generalized bias (for all parameters) with independent observations (data set 1)

Number of draws (R)	PMC	Standard Halton	Scrambled Halton	Shuffled Halton	Long shuffled Halton	Randomized shuffled Halton
5	6.333	9.803	1.900	4.742	6.686	4.380
50	0.451	0.190	0.156	0.310	0.472	0.524
100	0.286	0.223	0.132	0.389	0.275	0.208
200	0.224	0.153	0.167	0.167	0.158	0.154
500	0.174	0.151	0.144	0.162	0.267	0.161
1000	0.161	0.159	0.160	0.184	0.163	0.175
2000	0.196	0.163	0.161	0.180	0.158	0.139

100, this method yields a smaller generalized bias than PMC, shuffled Halton, or long shuffled Halton. The overall findings confirm conclusions from previous studies: that scrambled Halton sequences perform better than shuffled and other Halton sequences when nothing is amiss in the data series. In addition, it supports the notion that randomized shuffled Halton sequences yield fairly satisfactory results.

Table 8.2 shows the generalized bias values for correlated observations of data set 2. With this example, the six methods of MSLE sequence generation yield nearly the same results. This result suggests that concerns relating to observation correlation generated by QMCs may be unnecessary. However, it should be noted that the spatial correlation pattern inherent in data set 2 does not differ significantly from the patterns generated by standard, scrambled, and shuffled Halton sequences. The correlation between error terms of nearby observational units in space is always positive, paralleling the shuffled Halton's perfect positive correlation pattern. The rule for generating this synthetic data also suggests that the correlation peaks somewhere between every 30th and 50th observation pairing. These are multiples of 3 and 5, which were used as the bases for all five types of Halton sequence used. If the synthetic data set has a different pattern, and it is completely counter to the sequences, the pattern falsely imposed by sequences can obscure the real data correlation pattern and affect estimators, especially those associated with the variance-covariance matrix. Therefore, if Halton sequences must be used, a safe approach may involve shuffling the data prior to analyzing it so the potential correlation can be removed. Of course, if the order of observations is important to the model specifications, for example, in serial or spatial correlation analysis in which estimation methods are facilitated by proper ordering of observations.

Table 8.2 Generalized bias (for all parameters) with correlated observations (data set 2)

Number of draws (R)	PMC	Standard Halton	Scrambled Halton	Shuffled Halton	Long shuffled Halton	Randomized shuffled Halton
5	6.998	2.776	2.550	8.443	7.145	2.786
50	1.563	1.310	1.294	1.417	1.681	1.404
100	1.430	1.452	1.425	1.143	1.555	1.368
200	1.555	1.380	1.341	1.398	1.366	1.436
500	1.521	1.416	1.427	1.344	1.420	1.361
1000	1.396	1.384	1.400	1.446	1.327	1.468
2000	1.429	1.407	1.409	1.395	1.399	1.385

As to the performance of the randomized shuffled Halton sequence—thanks to the benefits of Halton sequence coverages—only when $R = 1000$ does PMC perform better. With all other draw numbers, the randomized shuffled Halton sequence appears to yield smaller errors. The results from this experiment suggest that, even when observations are correlated, QMCs and hybrid methods may be preferred to PMC because of their better coverage.

Table 8.3 shows estimates from just one of the 10 samples. For this sample, the estimates that result from assuming fixed coefficients on the *COST* and *REVENUE* terms are more accurate than those for the true random coefficients. These results also suggest that estimates from standard and scrambled Halton sequences are the most stable as R increases. With empirical data sets, in which true parameter values are not known, one signal for convergence, as Hensher and Greene (2003), Train (2003), and Walker (2001) have suggested, is the stabilization of estimate values. Therefore, this feature of standard and scrambled Halton sequences also makes them preferable for practice.

8.6 SUMMARY

With the growth of computational capability in recent years, simulation has been broadly adopted for complex model estimation. Significant effort has also been devoted to developing more efficient simulation techniques. This chapter discussed the advantages and potential problems of several simulation techniques, including standard, scrambled, shuffled, long shuffled Halton, and randomized shuffled Halton sequences. Standard sequences are found to generate periodic correlations across observations, with cycle lengths equal to their bases (or multiples of their bases, for higher dimensions). Though a scrambled Halton sequence can effectively disrupt a standard Halton's high correlation across alternatives, it cannot break the correlation across observations. A shuffled Halton sequence uses the same sequence of numbers for each observation, thus imposing perfect correlation across observations.

This issue of observation correlation deserves some attention, because in some spatially, temporally, or otherwise correlated data sets and models, there may be no alternatives for multidimensional integration. In these cases, there is not a need to consider shuffling or scrambling to break correlations across alternatives (in a model of discrete choices across alternatives). However, the interobservation correlation caused by a standard Halton sequence can be just as much of an issue. The correlation patterns across observations, when significantly different from the data set's true correlation patterns, can increase the simulation variance of the likelihood function and resulting estimates. For finite samples, this can contribute to a situation of empirical unidentifiability, wherein the true log likelihood function is already flat due to large sampling variance.

Table 8.3 Parameter estimates with different methods and data sets

Method	No. of draws (R)	Data set 1 (Independent observations)					Data set 2 (Correlated observations)				
	True value	β_{REV}	α_1	σ_1	α_2	σ_2	β_{REV}	α_1	σ_1	α_2	σ_2
		2	−1	10	1	5	2	−1	10	1	5
PMC	5	0.988	0.223	5.166	0.710	1.629	1.360	−0.390	3.283	0.848	1.603
	50	1.722	−0.975	9.408	0.888	4.986	1.990	−2.127	6.924	1.043	3.408
	100	1.737	−1.045	9.427	0.837	4.962	2.115	−2.253	7.218	1.096	3.606
	200	1.847	−1.136	9.917	0.781	5.074	2.108	−2.289	7.231	1.079	3.597
	500	1.828	−1.064	9.863	0.816	4.962	2.156	−2.177	7.355	1.092	3.643
	1000	1.818	−1.059	9.693	0.796	5.003	2.141	−2.152	7.120	1.077	3.624
	2000	1.824	−1.107	9.810	0.796	4.940	2.128	−2.145	7.123	1.090	3.605
Standard Halton	5	1.260	−1.475	9.350	0.536	3.406	1.728	−1.047	5.205	0.847	3.056
	50	1.815	−0.864	9.115	0.729	5.076	2.146	−2.096	7.070	1.066	3.684
	100	1.851	−1.120	9.931	0.787	5.013	2.112	−2.176	7.102	1.084	3.585
	200	1.835	−1.122	9.882	0.804	5.007	2.137	−2.102	7.023	1.085	3.665
	500	1.828	−1.099	9.813	0.804	4.964	2.135	−2.149	7.110	1.086	3.633
	1000	1.836	−1.123	9.903	0.811	4.985	2.133	−2.155	7.113	1.087	3.624
	2000	1.832	−1.114	9.853	0.811	4.980	2.133	−2.155	7.113	1.087	3.622
Scrambled Halton	5	1.392	−0.587	7.871	0.482	5.029	1.716	−1.202	5.275	0.846	3.421
	50	1.871	−1.048	9.689	0.770	5.223	2.108	−2.117	7.037	1.048	3.677
	100	1.820	−1.083	9.783	0.813	4.954	2.130	−2.141	7.115	1.085	3.657
	200	1.847	−1.115	9.945	0.808	5.043	2.126	−2.187	7.196	1.086	3.600
	500	1.838	−1.124	9.935	0.812	4.976	2.136	−2.138	7.071	1.086	3.642
	1000	1.830	−1.093	9.796	0.807	4.991	2.136	−2.164	7.139	1.089	3.621
	2000	1.833	−1.119	9.876	0.812	4.979	2.132	−2.156	7.116	1.087	3.622

Shuffled Halton	5	1.344	−3.118	12.545	0.804	3.505	1.863	−9.976	20.681	1.109	3.117
	50	1.813	−1.119	9.997	0.953	5.084	2.075	−2.364	7.639	1.236	3.493
	100	1.785	−0.965	9.550	0.820	4.996	2.104	−2.235	7.353	1.120	3.621
	200	1.820	−1.058	10.055	0.911	4.908	2.113	−2.059	7.034	1.141	3.636
	500	1.825	−1.110	9.972	0.843	4.939	2.129	−2.220	7.332	1.108	3.605
	1000	1.826	−1.104	9.932	0.847	4.959	2.128	−2.166	7.175	1.107	3.630
	2000	1.837	−1.219	10.198	0.847	4.953	2.137	−2.164	7.165	1.103	3.617
Long shuffled Halton	5	0.946	0.012	4.101	0.690	2.240	1.333	−0.654	3.473	0.904	1.421
	50	1.841	−0.797	9.197	0.614	5.447	2.087	−2.260	7.441	1.055	3.587
	100	1.834	−1.093	10.026	0.793	5.354	2.053	−2.104	7.028	1.076	3.515
	200	1.822	−1.044	9.476	0.770	4.997	2.126	−1.960	6.751	1.054	3.650
	500	1.805	−1.236	10.070	0.821	4.775	2.108	−2.249	7.290	1.098	3.535
	1000	1.823	−1.112	9.871	0.819	4.950	2.152	−2.171	7.127	1.096	3.653
	2000	1.837	−1.019	9.592	0.790	5.031	2.144	−2.163	7.131	1.094	3.634
Randomized shuffled Halton	5	1.152	−2.470	14.995	0.747	2.742	1.597	−1.786	6.586	0.891	2.183
	50	1.732	−1.788	11.769	0.884	4.764	2.078	−2.124	7.190	1.080	3.545
	100	1.793	−1.182	9.957	0.827	4.929	2.093	−2.108	7.081	1.070	3.556
	200	1.799	−1.101	9.687	0.781	5.026	2.114	−2.258	7.285	1.110	3.529
	500	1.823	−1.115	9.934	0.821	4.893	2.138	−2.180	7.114	1.094	3.612
	1000	1.828	−1.098	9.815	0.804	4.975	2.123	−2.165	7.103	1.082	3.605
	2000	1.834	−1.157	9.941	0.822	4.965	2.135	−2.140	7.113	1.090	3.615

In this work, a standard, scrambled, shuffled, long shuffled, and randomized shuffled Halton, together with PMC sequences, were applied to a synthetic data set in which random coefficients correlated across observations, using spatial processes. Fortunately, the results suggest that when the generic and imposed correlation patterns are not at odds, these QMC and hybrid methods are at least as good as PMC because they have better coverage.

Furthermore, the randomized shuffled Halton sequence was found to perform adequately with both independent and correlated observations. Considering that this method requires less computation time and memory than the scrambled sequence, the hybrid method can serve as a good alternative to standard and scrambled Halton sequences. It should be noted that many prior studies reveal that properties of a QMC sequence depend on its base, number of draws, and number of observations, as well as its initial position. Our knowledge of QMC is currently limited, so these newly-developed techniques should be used with caution. When the model is complicated and there is reason to believe potential observation correlation exists, PMC may be the safest choice.

Many opportunities for further study exist in this domain, including variations on interobservational dependence and comparisons of Bayesian and classical (MSLE) approaches for estimation of these complex models. As Bolduc et al. (1997) suggests, a Bayesian procedure requires approximately half as much computer time as MSLE with PMC. It would be interesting to see how Bayesian simulation techniques perform with correlated observations and how results compare to those based on QMC sequences for MSLE.

ACKNOWLEDGMENTS

This work is financially supported by the Benjamin H. Stevens Graduate Fellowship in Regional Science provided by the North American Regional Science Council. The authors are grateful to those who provided assistance for this study. These include Kenneth Train for his inspiring e-mails and Gauss code for estimating mixed logit models, Olivier Teytaud for his Matlab code-generating scrambled Halton sequences, Joan Walker for her e-mail replies, and Annette Perrone for her administrative and editing assistance.

ENDNOTES

[1] Integration error, or discrepancy, is the difference between true value (derived via integration) and the simulation-derived approximation.

[2] While Morokoff and Caflisch (1994) were the first to suggest a mathematically equivalent version of the shuffled Halton sequence, Hess and Polak (2003) developed their work independently and are the first to use the term shuffled Halton. They also were the first to apply this technique with a mixed logit model.

REFERENCES

Ben-Akiva, M., D. Bolduc, and J. Walker. 2001. Specification, estimation and identification of the logit kernel (or continuous mixed logit) model. Working Paper. Department of Civil Engineering, MIT. http://emlab.berkeley.edu/reprints/misc/multinomial2.pdf (accessed June 10, 2006).

Bhat, C. 2003. Simulation estimation of mixed discrete choice models using randomized and scrambled Halton sequences. *Transportation Research*, 37B(9): 837–855.

Braaten, E. and G. Weller. 1979. An improved low-discrepancy sequence for multidimensional quasi-Monte Carlo integration. *Journal of Computational Physics*, 33(2): 249–258.

Bolduc, D., B. Fortin, and S. Gordon. 1997. Multinomial probit estimation of spatially interdependent choices: An empirical comparison of two new techniques. *International Regional Science Review*, 20(1-2): 77–101.

Faure, H. 1992. Good permutations for extreme discrepancy. *Journal of Number Theory*, 42(1): 47–56.

Greene, W. 2002. *Econometric analysis*, 5th ed. Upper Saddle River, NJ: Prentice-Hall.

Halton, J. 1960. On the efficiency of evaluating certain quasi-random sequences of points in evaluating multi-dimensional integrals. *Numerische Mathematik*, 2: 84–90.

Hellekalek, P. 1984. Regularities in the distribution of special sequences. *Journal of Number Theory*, 18(1): 41–55.

Hensher, D. and W. Greene. 2003. The mixed logit model: The state of practice. *Transportation*, 30(2): 133–176.

Hess, S., and J. W. Polak. 2003. The shuffled Halton sequence. Working Paper. Centre for Transport Studies, Imperial College London. http://www.cts.cv.imperial.ac.uk/StaffPages/StephaneHess/papers/Hess_and_Polak_Shuffled_Halton_Sequences.pdf (accessed June 10, 2006).

McFadden, D. 1989. A method of simulated moments for estimation of the multinomial probit without numerical integration. *Econometrica*, 57(5): 995–1026.

McFadden, D. and K. Train. 2000. Mixed MNL models of discrete response. *Journal of Applied Econometrics*, 15(5): 447–470.

Morokoff, W. J. and R. E. Caflisch. 1994. Quasi-random sequences and their discrepancies. *SIAM Journal of Scientific Computing*, 15 (6): 1251–1279.

Munizaga, M. and R. Alvarez-Daziano. 2001. Mixed logit versus nested logit and probit. Working Paper. Department of Civil Engineering, The University of Chile. http://tamarugo.cec.uchile.cl/~dicidet/mmunizaga/mixed_logit.pdf (accessed June 10, 2006).

Niederreiter, H. 1992. Random number generation and quasi-Monte Carlo methods. Presented at CBMS-NSF Regional Conference Series in Applied Math, SIAM, Philadelphia, Pennsylvania.

Rouwendal, J. and E. Meijer. 2001. Preferences for housing, jobs, and commuting: A mixed logit analysis. *Journal of Regional Science*, 41(3): 475–505.

Sobol, I. M. 1967. On the distribution of points in a cube and the approximate evaluation of integrals. *U.S.S.R. Computational Mathematics and Mathematical Physics*, 7: 86–112.

Spanier, J. and E. H. Maize. 1994. Quasi-random methods for estimating integrals using relatively small samples. *SIAM Review*, 36(1): 18–44.

Srinivasan, K. and H. Mahmassani. 2005. Dynamic kernel logit model for the analysis of longitude discrete choice data: Properties and computational assessment. *Transportation Science*, 39(2): 160–181.

Train, K. 1999. Halton sequences for mixed logit. Technical Paper. Department of Economics, University of California, Berkeley. http://129.3.20.41/eps/em/papers/0012/0012002.pdf (accessed June 10, 2006).

——— 2003. *Discrete Choice Methods with Simulation*. New York: Cambridge University Press.

Walker, J. L. 2001. Extended discrete choice models: Integrated framework, flexible error structures, and latent variables. PhD Dissertation, Department of Civil and Environmental Engineering, MIT.

Wang, X. and K. M. Kockelman. 2006. Tracking land cover change in a mixed logit model: Recognizing temporal and spatial effects. Forthcoming in *Transportation Research Record*.

9

ANALYZING THE IMPACT OF LAND TRANSPORTATION ON REGIONAL TOURISM: THE CASE OF THE CLOSURE OF THE GLION TUNNEL IN THE VALAIS, SWITZERLAND

Miriam Scaglione
Institute for Economics & Tourism
HES-SO Valais University of Applied Sciences Valais–Switzerland

9.1 INTRODUCTION

The tourism sector has a significant role in the economy of the Valais. Its resorts, including Zermatt, Verbier, Crans Montana, Leukerbad, and Saas Fee, are well known throughout the world. Figures show the importance of tourism in the canton. In 2000, the share of jobs related to the sector was 18.7 percent, against an average of 9.8 percent for all Switzerland. The share of tourism (hospitality, restaurants, and transport) in the cantonal GDP of the Valais is 12.6 percent, twice the Swiss national level. The Valais does not have an international airport, only a military one at Sion, which occasionally allows a few charter flights. Therefore, land transport (by road and train) is the most important means of transportation affecting the rate of tourist arrivals.

The local topography is characterized by mountains and the Rhône River. Tunnels and passes are prevalent in the transportation network all around Switzerland.

The chain of the Alps stretches from Ventimiglia, near Nice, on the border between Italy and France at one end, to Vienna at the other end. It is a natural barrier that affects Switzerland, France, and Austria. In all, there are 14 routes through the Alps which are considered to be important. Specifically, through Switzerland there are four: the San Bernardino, linking the regional cities of Chur and Bellinzona; the Saint Gotthard, connecting Hamburg to Rome; the Grand Saint Bernard, linking Turin and Lausanne; and the Simplon, connecting Berne to Milan. The two latter transalpine routes pass through the Valais (Federal Office for the Development of Land Use, 2001).

One important route linking the Valais to the Geneva Airport is the national highway, which passes through the Glion Tunnel, a major access. This motorway is part of the transalpine route through the Grand Saint Bernard, linking Lausanne to Turin.

In both 2004 and 2005, the route was partially closed from April to November. The Glion tunnel consists of two separate tunnels of two lanes each, one for each direction. During the work, one was completely closed and traffic flowed through the other in both directions.

This chapter estimates the economic damage caused to the tourism sector by this closure. Nevertheless, the question: "How many more overnights would the Valais have had if the tunnel had not been closed?", and the same question regarding hotel revenues, cannot be answered because it presupposes a relationship of causality.

As an alternative, the goal of this research was to answer the same question but under the condition of *ceteris paribus* (other things being equal), that is, if all other conditions, such as the weather, the rate of exchange against the Swiss Franc, and all other variables remained the same. These latter conditions, though strong, are necessary to rule out any pretension of causality in the present study.

The research framework consists of finding a method that could link the dynamic of transportation variables to the dynamic of tourism variables. The analysis of their evolution over time seems to be a reasonable strategy to look for changes in those dynamics before and after the closure. Therefore, this research uses time series analysis methods not only as a forecasting tool but also as a tool to describe their evolution over time (*nowcasting*).

The two main sets of time series data relevant to tourism, hotel overnights and revenues, are available from the Office Fédéral de la Statistique (OFS) for the entire canton. Other Swiss federal and cantonal government departments—the Office Fédéral des Routes Suisses and Office Cantonal des Routes of the Valais—gather time series data from the automatic counting of traffic (that is, the number of vehicles) of the main tunnels and passes for the incoming tourism of the canton.

The choice of time series over other methods, such as econometric models, is explained by the following quotation from Andrew Harvey: "The distinguishing feature of a time series model, as opposed, say, to an econometric model, is that no attempt is made to formulate a behavioral relationship between y_t and other variables. The movements in y_t are 'explained' solely in terms of its own past, or by its position in relation to time" (Harvey, 1981). This latter characteristic of time series is central for the present research and is presented in a later section.

This research uses structural time series (STS) models and the Stamp (Version 6.0) software for STS (Koopman, Harvey, Doornik, and Shepard, 2000). The STS model captures the salient characteristics of stochastic phenomena, usually in the form of trends, seasonal or other irregular components, explanatory variables, and intervention variables, which greatly contributes to thorough comprehension of the phenomena. If a STS model has explanatory variables as a component, it is referred to as a *causal model* and the interpretation of those coefficient variables is the same as in econometric models; otherwise, if the model lacks explanatory variables, it is referred to as a *no-causal* one.

Moreover, STS can be univariate or multivariate, depending on the type of the endogenous variable; in the latter case this variable is a vector. The choice of the multivariate model is appropriate when the variables belong to the same economic system. These models are called seemingly unrelated time series equations (SUTSE) and are an extension of univariate forms with the advantage of allowing for cross-correlation leads between variables. In Stamp, SUTSE are particularly attractive because, on one hand, models with common factors emerge as a special case, on the other, the direct analysis of the unobservable components provides a more efficient forecast and inference (see Appendix 1 for details).

The following sections present each step of the research framework:

1. Analyzing the data and filling the missing value gaps using univariate models.
2. Modeling a SUTSE (bivariate), having a vector composed of overnights and revenue as the endogenous variable, and transportation series data as the endogenous variables.
3. Identifying the coefficient of the significant exogenous variables and those that show structural breaks during the closure period, that is, those coefficients whose estimates do not remain the same during that time.

4. Using a univariate model, forecasting the exogenous variables that show structural breaks in the previous step during the closure time. This provides the estimates of the traffic if the tunnel had not been closed, *ceteris paribus*.
5. Forecasting the endogenous tourism variables using the SUTSE model as in Step 2, but after replacing the explanatory variables by the forecast obtained in the previous step during the closure time. This step gives the estimate of tourism variables if the tunnel had not been closed, *ceteris paribus*.
6. Calculating the difference between forecast and observed values of the tourism variables.

Steps 1 and 3 are the *nowcasting* part of the framework, whereas the following ones are its forecasting part.

9.2 DATA DESCRIPTION AND ESTIMATION OF MISSING VALUES

The research is based on the hypothesis that there is covariance between transportation variables and tourism ones. Therefore, tourism variables are considered endogenous or dependent, whereas transportation variables are exogenous or independent ones.

9.2.1 Endogenous Variables

The OFS of Switzerland supplies the time series of tourism variables. As noted, this research uses two monthly series: overnights and revenue for the hotel sector. The data do not contain the values of other hospitality resources such as camp sites, sanatoria, dormitories for groups, youth hostels, and rented apartments and villas.

The period of the sample is from February 1992 to September 2005, with a gap of missing values for the whole of 2004. During that year, the OFS was re-engineering its tourism department and ceased to gather the necessary data. Therefore, the cantonal government decided to gather the data during 2004 itself, through Valais-Tourisme. Unfortunately, it collected only overnights series and not revenues from hotels.

Given that the first partial closure of the Glion tunnel took place from April to November 2004, the estimation of revenues during that period was indispensable. Also, the closure was an exceptional event and the interpolation using a no-causal univariate model would not be able to handle this latter fact.

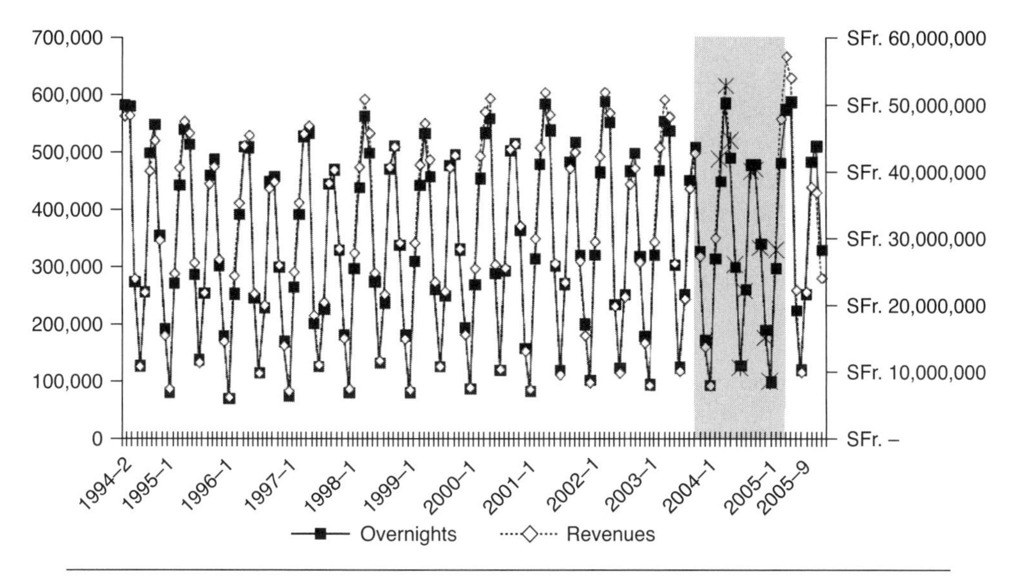

Figure 9.1 Revenue (right *y*-axis) and overnights (left *y*-axis) time series overall in the Valais. Estimate revenue observations in 2004 represented by a star in the gray band.

Therefore, the author analyzed the covariance of the overnights with revenue series using a no-causal bivariate (SUTSE) model from February 1994 to December 2003 having an endogenous variable as the vector $\binom{\text{overnights}_t}{\text{revenue}_t}$ in log. The diagnostic of goodness-of-fit of the model is not included here but no aspect of the diagnostic test was significant. The most important information is that the cross-correlation between the two series (overnights and revenue) gives 0.95.

Thus, a good solution for filling the gap of missing values is to build a causal model having as endogenous variable *revenue* and as exogenous variables the *overnights*, all in log. The diagnostic of goodness-of-fit is based on the following statistics: s = 2.1E-2, $N = 6.46$, $H(35) = 0.76$, $DW = 1.72$, $Q(6) = 12.35$, $R^2 = 0.92$, which shows that none are significant. The elasticity coefficient for overnights was also 0.92. Figure 9.1 shows the final result of the process.

9.2.2 Exogenous Variables

Exogenous variables are automatic counts of vehicles on the main incoming routes into the canton. Unfortunately, the automatic counts did not function during some months, therefore, some values are missing. The sample of data begins February 1994 because the January 1994 count is missing for Glion. Table 9.1 gives the period of missing values for each counting position.

To fill the gaps of missing values, the author used two techniques, depending on the size of the gap. If it were less than four months, she calculated a model for each gap using univariate and no-causal models, forecasting two periods more than the size. The value retained was selected from the estimate, as were the two limits of the confidence interval (± standard deviation). The criterion retained was used to choose the value whose mean square error for the two remaining known values was the minimum. When the size of the gap was greater than three months, the method used was similar to the one explained for the endogenous revenue variable: the author used forecasting and prediction graphics routines in an iterative manner.

The present study takes into account data from seven counting stations on different routes; the following paragraphs present a description of the geographical situation and the importance of each of them.

The A9 highway through Glion is a main link between the regions around the Lake of Geneva, but it also carries a flow of traffic coming from the German-speaking part of Switzerland and the Rhône Valley. Moreover, this axis is used by international traffic passing through the Grand Saint Bernard and the Simplon passes.

An alternative route to the A9 is the cantonal road, Vevey-Lausanne (K9), which crosses through the city of Montreux. Less important, but also used as

Table 9.1 Dates and number of missing values by routes

Variable (access routes to the Valais)	Missing values dates	Total months
Glion (A9)	May to July '94 August to September '95 October '98 February to December '99 August to September '00 December '00 August to September '01 December '01 January '02	23
Vevey-Lausanne (K9)	June '02	1
Grand St. Bernard tunnel	nil	0
Kandersteg-Frutigen	January '01	2
Simplon tunnel	nil	0
St. Bernard pass	June to July '97 April to May '00	4
Vionnaz-St.-Gingolph	nil	0

alternatives to the A9, are the Col des Mosses and Col du Pillon passes; however, these two routes are not taken into account as exogenous variables in the present study. All these substitute routes add transalpine capacity to the flow toward the Valais, not only for tourism purposes but also regional and local traffic between the Chablais and the Riviera of the neighboring canton of the Valais, Vaud.

The Grand Saint Bernard pass and tunnel are alternative routes to the Valais along the north–south (Switzerland–Italy) axis. The counting machine is located on the road leading not only to the Grand Saint Bernard tunnel but also to the Val de Bagnes Valley and the well-known resort of Verbier. In March 1999, the Grand Saint Bernard tunnel was closed due to a fire and the pass became the main substitution route for that axis.

The counting station in the Kandertal between Frutigen and Kandergrund monitors the traffic toward the Alpine resort of Kandersteg in the neighboring canton of Bern. Nevertheless, a great part of the traffic is related to rail transportation through the Lötschberg tunnel.

The Simplon Pass is one of the main Swiss north–south axis through the Alps, the summit being at an altitude of 2,000 meters. The Simplon route is not only more sensitive to weather conditions than the others, but also the longest. The Grand Saint Bernard tunnel or pass are alternatives to this route. Moreover, the motorway has not been completed between the cities in the Valais of Sierre and Brig at the foot of the Simplon pass.

Saint-Gingolph is a town near the border between Switzerland and France and is the entrance to the Valais when passing along the southern side of the Lake of Geneva. This route is far less important than K9. It carries only a small part of the traffic between Geneva and the Valais, and people working in Switzerland but living in France are the main users. This axis could increase in importance when the road work is undertaken to replace the bottleneck at the Pont du Scex. Moreover, discussions are taking place to evaluate an extension of the highway from Annemasse to Thonon-les-Bains (both in France) toward Evian (France) and Saint-Gingolph.

9.3 ANALYZING THE TOURISM AND TRANSPORTATION DYNAMIC (NOWCASTING)

A SUTSE model has the vector $\begin{pmatrix} \text{overnights}_t \\ \text{revenue}_t \end{pmatrix}$ as an endogenous variable and transportation data as the exogenous variables—counts of vehicles—all in log. The sample taken into consideration covers the period February 1994 to December 2003.

Equation (9.1) shows the model in a schematic manner (the formal definition is in Appendix 1):

$$\begin{pmatrix} \text{overnights} \\ \text{revenue} \end{pmatrix}_t = \begin{pmatrix} \text{Trend}^{overnights} \\ \text{Trend}^{revenue} \end{pmatrix}_t = \begin{pmatrix} \text{seasonal}^{overnights} \\ \text{seasonal}^{revenue} \end{pmatrix}_t + \sum_{\substack{all \\ routes}} \begin{pmatrix} \beta_{route}^{overnights} \\ \beta_{route}^{revenue} \end{pmatrix} * \text{route}_t + \begin{pmatrix} \varepsilon^{overnights} \\ \varepsilon^{revenue} \end{pmatrix}_t \tag{9.1}$$

For the overnights component, the equation standard error is 5.6E-2 and the diagnostic of goodness-of-fit is based on the following statistics: $N = 0.43$, $H(35) = 0.78$, $DW = 1.97$, $Q(7) = 6.01$, $R^2 = 0.56$. None of them is significant. The same thing happens for the revenue component: its equation standard error is 5.5E-2 and the diagnostic data of goodness-of-fit are $N = 3.6$, $H(35) = 0.87$, $DW = 2.0$, $Q(7) = 3.41$, $R^2 = 0.51$.

The R^2 coefficient can appear low if the reader gives the same interpretation as in ordinary least square (OLS), but in the case of STS, and when the series is not stationary, the coefficient is calculated differently (the interpretation is explained in Appendix 1).

Table 9.2 shows the elasticity coefficients. For the Grand Saint Bernard pass it is not significant; furthermore, its sign is negative. This is in line with the findings of an intercantonal working group (of the cantons of Vaud and Valais) created to analyze the evolution of traffic during the closure. The Saint Bernard pass road is used mainly by excursionists who choose it to admire the Alpine landscapes. Their analysis is based on the study of the daily counts per hour (personal communication with M. Putallaz, Office Cantonal des Routes of the Valais).

The case of Saint-Gingolph is somewhat different. Though people familiar with the area consider the route mainly important for tourists having a second home and workers employed outside the Valais, in the analysis its elasticity coefficient is, for some periods, slightly positive.

The Kandersteg-Frutigen route, however, is not significant. The explanation may be that it is used mainly by tourists having second homes in the Valais but living in the German-speaking part of Switzerland. Nevertheless, further studies which analyze the overnight values by origin—Swiss or foreign—are warranted; this, however, is beyond the scope of the present research.

The author decided to recalculate the model, excluding the Saint Bernard pass and using an estimated sample from February 1994 to December 2003. For the overnights component, the equation standard error is 5.6E-2 and the diagnostic of goodness-of-fit is based on the following statistics: $N = 0.41$, $H(35) = 0.78$, $DW = 1.98$, $Q(7) = 6.02$, $R^2 = 0.56$. None of these is significant. The same occurs for the revenue component: its equation standard error is 5.6E-2 and the diagnostic data of goodness-of-fit data are: $N = 3.9$, $H(35) = 0.87$, $DW = 2.0$, $Q(7) = 3.12$,

Table 9.2 Elasticity coefficient for overnights and revenue (standard deviation in brackets)

Variable (access routes to the Valais)	Overnights		Revenue	
	Including Gr St Bernard pass	Excluding Gr St Bernard pass	Including Gr St Bernard pass	Ecluding Gr St Bernard pass
Glion (A9)	0.51** (0.21)	0.48** (0.21)	0.44** (0.21)	0.41** (0.21)
Vevey-Lausanne (K9)	0.29* (0.17)	0.28* (0.17)	0.16 (0.17)	0.15 (0.16)
Grand St. Bernard tunnel	0.21** (0.07)	0.19** (0.06)	0.18** (0.07)	0.15** (0.62)
Kandersteg-Frutigen	0.12 (0.07)	0.05 (0.10)	0.11 (0.11)	0.06 (0.096)
Simplon tunnel	0.11** (0.05)	0.10** (0.02)	0.1** (0.05)	0.095** (0.045)
Grand St. Bernard pass	−0.14 0.21)	nil	−0.18 (0.21)	nil
St.-Gingolph	−0.25 (0.26)	−0.25 (0.2)	−0.1 (0.26)	−0.1 (0.26)

$R^2 = 0.50$. These figures are nearly the same as those for the preceding models, including the Saint Bernard pass.

In order to evaluate the changes in the relative importance of the routes during the closure, the author analyzed the final state estimate produced by Stamp, which contains not only all the information needed for the forecast, but also the regression estimates of the explanatory variables.

Therefore, the author ran the model in an iterative manner, each time increasing the upper limit of the sample by one observation forward, that is, from January 2004 to September 2005, the last observation of the transportation database—22 times in all. Figure 9. 2 shows the elasticity coefficients obtained at each iteration for the overnights and revenue components.

It is clear that the two most significant changes are recorded by Glion (A9) and Lausanne-Vevey (K9), that is, in the interaction between the national highway and the cantonal road. Further analysis of the *t*-values shows that, during the first closure of 2004, the elasticity coefficient of Glion was significant at the average

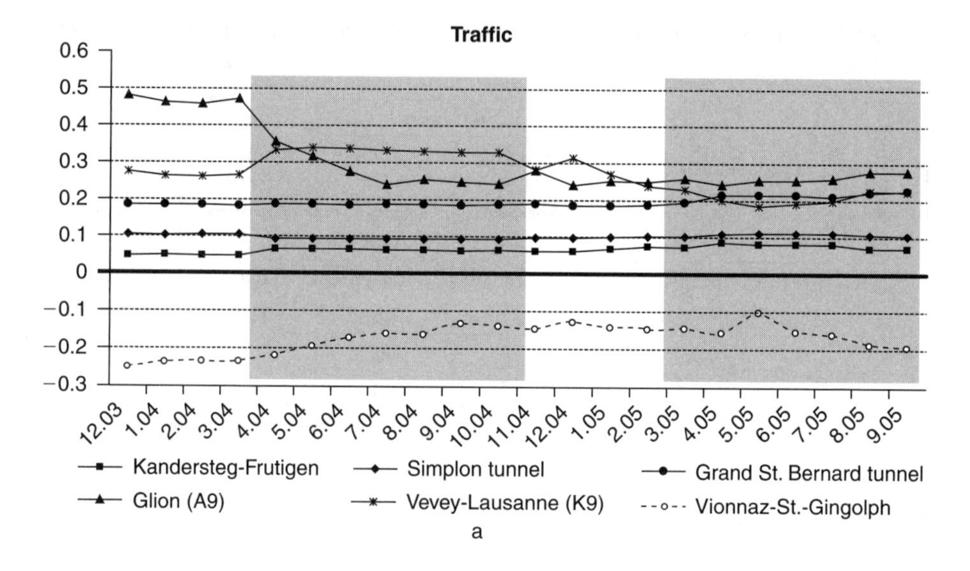

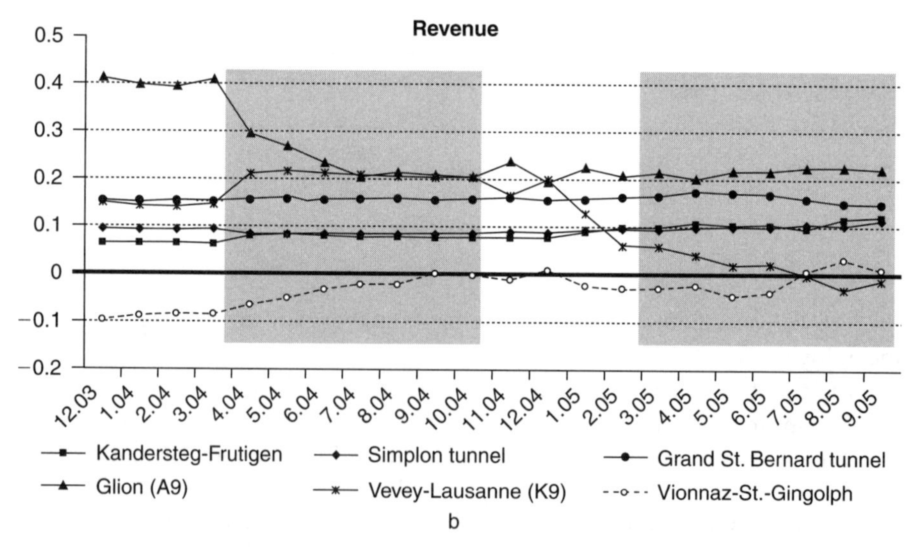

Figure 9.2 Elasticity coefficients for transport variables; the grey band shows the closure period. Axis *x* shows the last observation included in the sample. (a) Overnights elasticity coefficients, (b) revenues ones.

level of 10 percent, whereas in the second closure period of 2005, the level of significance was on average 5 percent. The opposite, more or less, occurs to the cantonal road (K9), which is significant at an average level of 5 percent, but not significant during the second closure at a level of 10 percent.

Figure 9.3 shows that during the first closure, tourist hotel customers preferred the Lausanne-Vevey (K9) route over Glion, but in the second closure the opposite occurred. This is in line with the facts observed by the Office Cantonal des Routes of the Valais: the rate of flow at Glion was maximal during the second closure.

The conclusion is that the two routes most affected by the closure of the tunnel were Glion (A9) and Lausanne-Vevey (K9).

9.3.1 Forecast of Glion (A9) and Lausanne-Vevey (K9)

As noted, the time series method allows for explaining and modeling of a series solely in terms of its own past, as does STS. The goal of this section is to find the figures of the transport variables if the tunnel had not been closed, *ceteris paribus*. With this in mind, to forecast those figures, an univariate and no-causal model seems to be an appropriate strategy. Both models are calculated in logs on the February 1994 to March 2004 sample. The values from April 2004 to September 2005 are forecasts.

The model for Glion (A9), the equation standard error 2.8E-2, and the diagnostic of goodness-of-fit are based on the following statistics (see Appendix 1): $N = 1.56$, $H(36) = 0.74$, $DW = 1.75$, $Q(7) = 5.27$, $R^2 = 0.41$. A significant ($p < 0.001$) level intervention in February 2002 is present in the model. Beside the slight, low value for DW statistics, none of the other tests is significant. Figure 9.3 shows forecast and observed values. As expected, the forecast overestimated observed values during the closure period of the Glion tunnel.

Regarding the model built for Lausanne-Vevey (K9), its equation standard error is 2.9E-2 and the diagnostic data of goodness-of-fit are $N = 0.03$, $H(36) = 0.94$, $DW = 1.86$, $Q(7) = 6.93$, $R^2 = 0.59$; it has an impulse intervention in May

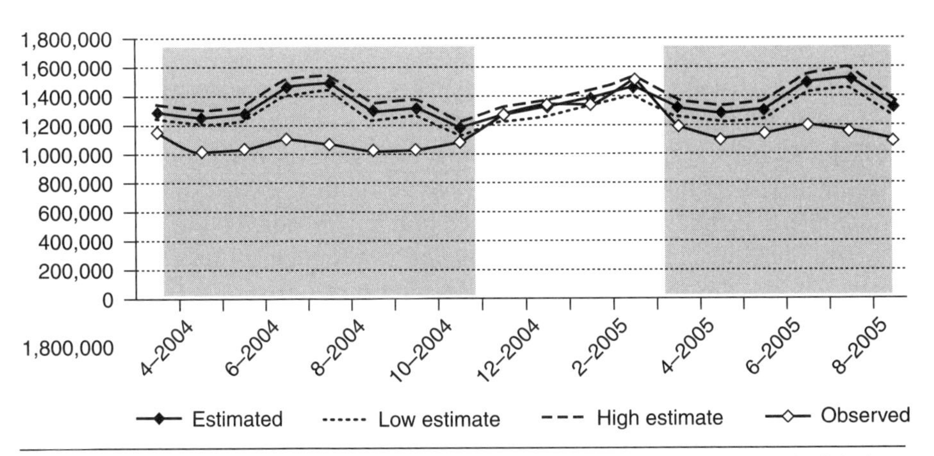

Figure 9.3 Forecast and observed values for Glion, confidence interval ±one standard deviation. Grey bands show closure period.

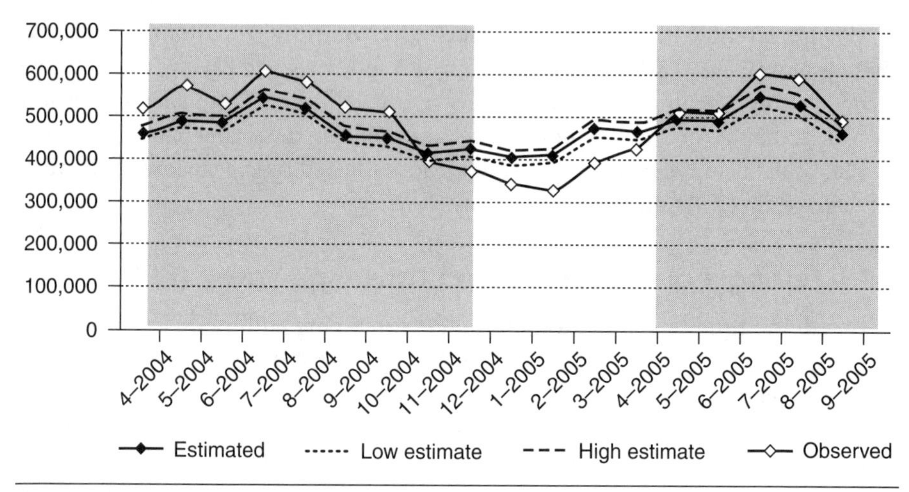

Figure 9.4 Forecasted and observed values for Lausanne-Vevey (K9), confidence interval ±one standard deviation. Grey bands show closure periods.

1994, but this value is an estimate because it was missing in the original database (see Figure 9.1). Figure 9.4 shows forecast and observed values for the cantonal road. As expected, forecast values underestimated observed ones during the closure time.

The next section presents the calculation of two different forecasting methods to evaluate overnights and revenue figures during the closure and based on the estimates obtained in this section.

9.3.2 Forecasting the Endogenous Tourism Variables

The series for Glion (A9) and Lausanne-Vevey (K9) are updated with the values obtained in the previous section from April 2004 to September 2005 as exogenous variables. Other transportation series remain the same as in Section 9.2. In order to estimate endogenous tourism variables based on this new database, we use the two different methods and routines: prediction test and forecasting.

The values obtained using the prediction test routine are referred to as *estimates*, whereas the ones obtained by forecasting are referred to as *forecasts*.

Figure 9.5 shows estimates and forecast values for overnights (see Figure 9.5a) and revenues (see Figure 9.5b). Figures obtained using both methods are similar and are both contained in the interval of ± standard deviation of the estimates.

Figure 9.5b shows the revenue figures only for 2005 because the values for 2004 were not available and were estimated in Section 9.1. The post-predicted tests of the prediction test routine are not significant, showing that there is no

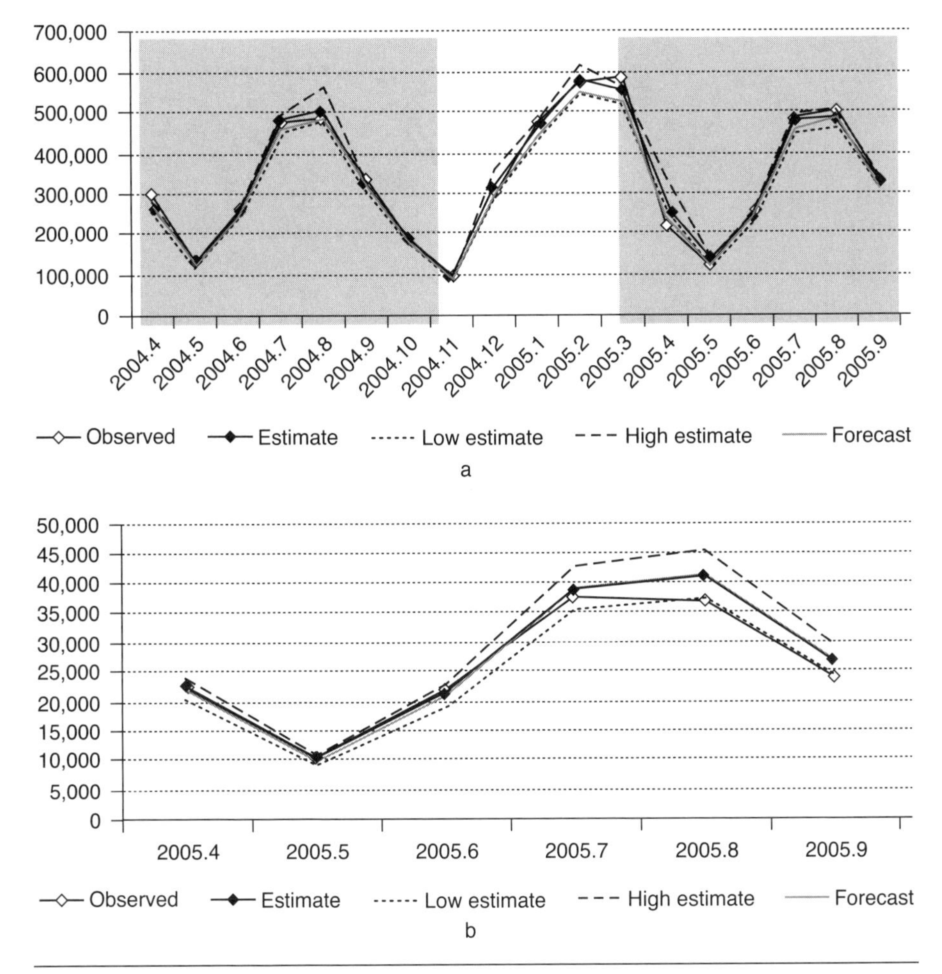

Figure 9.5 Estimate, forecast, and confidence interval ±one standard deviation. (a) overnights, grey bands show closure periods, (b) revenue in millions of SFr.

structural break in the model. There is, therefore, not enough evidence that the elasticity coefficient of the model in Section 9.2 changes in the estimate sample of April 2004 to September 2005.

9.3.3 Estimates of Figures Lost

Figure 9.6 shows the percentage between estimates and forecast values with observed variables. In 2004, for overnights, the only month that shows a negative difference between estimates or forecasts and the values observed is August 2004; the difference with the estimated value is 25,647 overnights (5.4 percent)

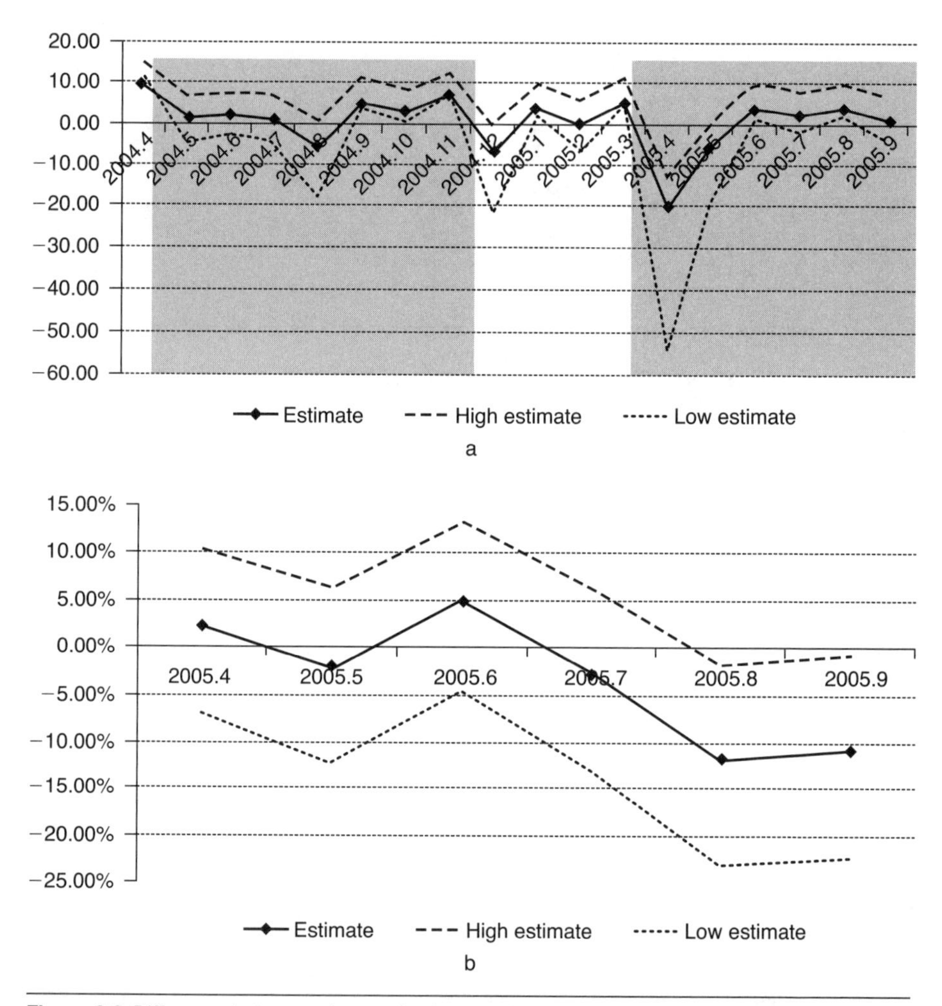

Figure 9.6 Difference between observed values and estimates, forecasts. Confidence interval ±one standard deviation plotted in grey. (a) overnights, grey bands show closure periods, (b) revenue second closure.

lost. The fact that the difference for December 2004 is also negative, 6.6 percent (19,548 overnights), is because the closure up to the preceding month must be confirmed by using other data, such as those obtained from surveys.

The overnights lost in 2005 are greater, especially in May 2005, which lost 20.3 percent (44,896): the figure of June 2005 is less than this, at 5.7 percent (6,755 overnights). Both bands of the confidence interval for each figure are also negative, giving enough evidence that there was a loss of overnights during those months.

These findings are in line with the elasticity coefficient of transportation variables (see Figure 9.2a). Glion (A9) has only a moderate value during the closure, whereas the elasticity of Lausanne-Vevey (K9) is smaller. Therefore, the slight recovery of Glion (A9) in 2005 seems to be insufficient to compensate the hospitality industry for the decrease on the Lausanne-Vevey (K9) route.

All these negative differences total −96,845 overnights, whereas all the differences taken together (both negative and positive) give a figure of 219,634, which leaves a positive overall balance for the two closure periods. Nevertheless, the analysis carried out on the closure period of 2005 produces a negative balance of 7,534 overnights.

For revenue, the analysis is restricted to 2005 because figures for 2004 have been estimated due to the lack of data. For May 2005, the difference between the forecast and the observed value of revenue is −2.5 percent (−268,000 SFr). In July 2005, the difference is −3.3 percent (−1.225 million SFr). The most significant differences were found during August 2005 and September 2005; moreover, the limit of ±standard deviation is also negative, giving sufficient evidence for loss of revenues. In August 2005, the difference is −12.2 percent (−4.484 million SFr) and in September 2005, −11.3 percent (−2.703 million SFr).

The addition of all those negative differences figures is −8.680 million SFr, whereas the total of the whole balance (negatives and positives) is −7.285 millions, that is, still negative.

9.4 SUMMARY

The losses in tourism figures during the second closure, *ceteris paribus*, show that tourists changed their behavior in comparison with their reaction during the first closure. It seems that experience was stronger than communication. During the two closures there was a large campaign to distribute information through various channels, that is, through a dedicated Website and on the radio. Moreover, national radio channels gave announcements of the delay expected at the tunnel in both directions every hour. Nevertheless, the experience of delay during the first closure, and probably word-of-mouth as a result, seem to have been stronger than information. The only improvement that these efforts seemed to bring about was the fact that Glion (A9) slightly increased its importance during the second closure.

The other significant fact was that, although Glion reached the same as the univariate forecast values (Figure 9.3) between the two closures, Figure 9.2a does not reflect it. This could show that hotel customers are more sensitive to transportation problems than are other categories, that is, owners of second homes or those going to clinics/sanatoria, rented apartments, or staying in villas.

An analysis of rail traffic frequency from June 1999 to May 2005, not included in this chapter, shows that there is not enough evidence to suggest a transfer from road to rail transport in the case of hotel customers. The cross-covariance of the trend of overnights and rail traffic is 3.1E-2. There is, nevertheless, a significant increase in traffic in April 2004, 20.7 percent, and in July 2004, 37.7 percent. To explain the poor correlation with hotel overnights, the author studied the train traffic series on a daily basis. The seasonal analysis shows that there are only two days of the week as a high season, Friday and Saturday; all others being a low season. This suggests that train travel is used mainly by excursionists or by people having second homes in the Valais but not by hotel customers. Nevertheless, further studies are needed to confirm this hypothesis.

The application of time series methods to public and easily-available data can provide an estimate of hypothetical figures, *ceteris paribus*, and also certain insights into changes in transport behavior. This method is less expensive and often more accurate than surveys, especially when the latter are not randomly sampled. Moreover, the bias due to automatic counting can be estimated because it is more likely to be known and corrected.

9.5 APPENDIX 1

The aim of the STS model is to capture the salient characteristics of stochastic phenomena, usually in the form of trends, seasonal or other irregular components, explanatory variables, and intervention variables. This model can reveal the components of a series which would otherwise be unobserved, greatly contributing to thorough comprehension of the phenomena. This description describes only the elements necessary for this study; for a complete description see Harvey, 1990 and Koopman, Harvey, Doornik, & Shephard, 2000. A STS multivariate model may be specified as:

$$\textit{Observed variables} = \textit{trend} + \textit{cycle} + \textit{intervention} + \textit{irregular}$$

The algebraic form for N series is:

$$y_t = \mu_t + \psi_t + \Lambda I_t + \varepsilon_t \quad \varepsilon_t \sim NID(0, \Sigma_\varepsilon^2) \quad t = 1, \cdots, T \tag{9.1}$$

Unless otherwise stated, the elements in Equation (9.1) are (Nx1) vectors. y_t is the vector of observed variables, μ_t is the stochastic trend, ψ_t. is the cycle, denotes the NxK* matrix of coefficients for the interventions and I_t the $K^* \times 1$ vector of interventions. The stochastic trend is intended to capture the long-trend movements in the series and trends other than linear ones, and is composed of two elements:

the levels (2) and the slopes (3). The trend described allows the model to handle these:

$$\mu_t = \mu_{t-1} + \beta_{t-1} + \eta_t \quad \eta_t \sim NID\left(0, \Sigma_\eta^2\right) \quad t = 1, \cdots T \tag{9.2}$$

$$\beta_t = \beta_{t-1} + \varsigma_t \quad \varsigma_t \sim NID\left(0, \Sigma_\varsigma^2\right) \quad t = 1, \cdots, T \tag{9.3}$$

If the variances of the irregular components ε_t in Equation (9.1), the disturbances of the level η_t in Equation (9.2) and at least one of the slope terms ζ_t are simultaneously strictly positive, the model is called a local linear trend (LLT).

When the level component is fixed and different for null, and when the two other variances are not null, the model is called a smoothed trend with a drift.

For an univariate model, the cycle ψ_t has the following statistical specification:

$$\begin{bmatrix} \psi_t \\ \psi_t^* \end{bmatrix} = \rho \begin{bmatrix} \cos\lambda_c & \sin\lambda_c \\ -\sin\lambda_c & \cos\lambda_c \end{bmatrix} \begin{bmatrix} \psi_{t-1} \\ \psi_{t-1}^* \end{bmatrix} + \begin{bmatrix} \kappa_t \\ \kappa_{t-1} \end{bmatrix}, \quad t = 1, \cdots T \tag{9.4}$$

where λ_c is the frequency, in radians, in the range $0 < \lambda_c < 1$, κ_t and κ_{t-1}^*, are two mutually uncorrelated white noise disturbances with zero means and common variance σ_κ^2, and ρ is the damping factor. The period of the cycle is $2\pi/\lambda_c$.

Cycles can also be introduced into a multivariate model. The disturbances may be correlated; the same, incidentally, can occur with any components in multivariate models. Since the cycle in each series is driven by two disturbances, there are two sets of disturbances and Stamp assumes that they have the same variance matrix (Koopman et al., 2000), that is:

$$E(\kappa_t \kappa_t') = E(\kappa_t^* \kappa_t^{*\prime}) = \Sigma_\kappa, \quad E(\kappa_t \kappa_t^{*\prime}) = 0, t = 1, \ldots, T \tag{9.5}$$

where Σ_κ is a $N \times N$ variance matrix.

Stamp has pre-programmed the following exogenous intervention variables used in this study:

1. AO: it is an unusually large value of the *irregular* disturbance at a particular time. It can be captured by an *impulse* intervention variable which takes the value of the outliers one at the time and zero elsewhere. If t_{ao} is the time of the outlier, then the exogenous intervention variable $I_{t_{ao}}$ has the following form:

$$I_{t_{ao}} = \begin{cases} 1 \text{ if } t = t_{ao} \\ 0 \text{ if } t \neq t_{ao} \end{cases} \quad t = 1, \ldots, T$$

2. LS: this kind of intervention handles a *structural break* in which the level of the series shifts up or down. It is modeled by a *step*

intervention variable which is zero before the event and 1 after it. If t_{LS} is the time of the level shift, then the exogenous intervention variable $\mathrm{I}_{t_{LS}}$ has the following form.

$$\mathrm{I}_{t_{LS}} = \begin{cases} 0 \text{ if } t > t_{LS} \\ 1 \text{ if } t \geq t_{LS} \end{cases} \quad t = 1, \ldots, T$$

Diagnostics

The diagnosis test statistics for a single series in an STS model are (see Koopman, Harvey, et al., 2000; Harvey, 1990):

- *s*: The equation standard error
- *N*: the Doornik-Hansen statistic, which is the Bowman-Shenton statistic with the correction of Doornik and Hansen. Under the null hypothesis that the residuals are normally distributed, the 5 percent critical value is approximately 6.0.
- *H*(df): A two-sided *F*-test that compares the residual sums of squares for the first and last thirds of the residuals series.
- *DW*: or Durbin Watson test statistic, distributed approximately as $N(0,1/T)$, T being the number of observations.
- *Q*(df): Box-Ljung *Q*-statistic, test of residual serial correlation, based on the first P residual autocorrelations and distributed as chi-square with $P-n+1$ df, when n parameters are estimated.
- R^2: Coefficient of determination. Stamp gives three different evaluations of the coefficient of determination depending on the kind of model adjusted. If the series is stationary with no trend and seasonal, the coefficient of R^2 is the same as OLS models.

If the series shows trend movements, the coefficient (R_D^2) is based on the comparison of prediction error variance on the first difference $\Delta y_t = y_t - y_{t-1}$ as follows:

$$R_D^2 = 1 - \frac{SSE}{\sum_{T=2}^{T} (\Delta y_t - mean(\Delta y))^2}$$

where SSE stands for sum of square errors.

The interpretation of such coefficient is minimally different from the usual R^2 in OLS. In this case, this coefficient measures the reduction of variance in comparison with an OLS model.

If the series shows not only trend but also seasonal pattern, the coefficient (R_S^2) which measures the reduction of variance in comparison with a model having a random walk plus a drift and fixed seasonal components. This latter coefficient has the following form:

$$R_S^2 = 1 - \frac{SSE}{SSDSM}$$

where SSDSM stands for the sum of the square of first differences around the seasonal means.

ACKNOWLEDGMENTS

This research project was funded thanks to a grant from the Department of the Economy and Territory, of the Secretariat for the Economy and Tourism of the Canton of Valais, Switzerland. The author wishes to thank the following persons for their kind assistance in the preparation of this chapter: Mr. François Seppey (Chief of the Economy and Tourism Secretariat of the Canton du Valais), Professor Dr. Peter Keller (former Director for Tourism, Secretariat of State for Economic Affairs, SECO, Swiss Federal Government, and Head of the Tourism Institute at the School of Business Studies (HEC) of the University of Lausanne), Mr. Peter Hugg (Swiss Federal Railways, CFF), Mr. Jean-Michel Putallaz (Head of the Office Cantonal des Routes, Valais). The author is particularly indebted to Professor Dr. Martin Schuler (Federal Polytechnic Institute, EPFL, Lausanne, Switzerland), who provided all the explanations regarding the geographical situation and the importance of every counting station. Mr. Merrick Fall was responsible for editing the English text. Any possible inaccuracy which the text may contain is the sole responsibility of the author.

REFERENCES

Federal Office for Spatial Development. 2001. *Through the Alps. Transalpine freight traffic by road and rail.* Bern: Federal Department of the Environment, Transport, Energy and Communication.

Harvey, A. C. 1981. *Times Series Models.* Oxford, UK: Philip Allan; Atlantic Highlands, NJ: Humanities Press.

——— 1990. *Forecasting, Structural Time Models and Kalman Filter.* Cambridge, UK: Cambridge University Press.

Koopman, S. J., A. C. Harvey, J. A. Doornik, and N. Shepard. 2000. *Stamp.* London: Timberlake Consultant.

10

QUASI-LIKELIHOOD GENERALIZED LINEAR REGRESSION ANALYSIS OF FATALITY RISK DATA

C. Craig Morris
Bureau of Transportation Statistics
Research and Innovative Technology Administration
U.S. Department of Transportation

10.1 INTRODUCTION

Transportation-related fatality risk is a function of many interacting human, vehicle, and environmental factors. Statistically valid analysis of such data is challenged both by the complexity of plausible structural models relating fatality rates to explanatory variables and by the uncertainty regarding the probability distribution of the data. But, fortunately, generalized linear modeling and maximum quasi-likelihood estimation together provide an extraordinarily effective set of statistical tools for the analysis of such data. The goal of this chapter is to illustrate and promote the application of these tools to fatality risk analysis.

First, we review the basic principles of generalized linear modeling and quasi-likelihood estimation. Next we illustrate the use of these tools to analyze the association of motorcyclist fatality rates with U.S.–state helmet laws while controlling for climate measures (annual heating degree days and precipitation inches) as proxies for motorcycling activity. The focus of the illustration is on Poisson log-linear regression analysis of over-dispersed fatality rate data via quasi-likelihood generalized linear modeling methods.

One popular reference on generalized linear modeling and maximum quasi-likelihood is McCullagh and Nelder's (1989) *Generalized Linear Models.* Another is Agresti's (2002) *Categorical Data Analysis.* Together, these venerable resources provide the essential theoretical and practical background for a variety of multivariate analyses on either continuous or categorical data, including analysis of variance and multiple linear regression, logistic, logit, probit, or log-linear regression, among others. What remains for application of these methods is the choice of statistical software (although one could implement the methods from scratch), and there are many choices. The example we review later in this chapter uses the SAS (2005) V8.0 procedure GENMOD (for generalized linear modeling). Methods for longitudinal, repeated-measures, spatially-correlated, or other correlated data are also available, but these are beyond the scope of this introductory tutorial; two fine reference texts for correlated data problems are Diggle et al. (2002) *Analysis of Longitudinal Data* and McCulloch and Searle (2001) *Generalized, Linear, and Mixed Models.*

10.1.1 Generalized Linear Models

Generalized linear models are an extension of *classical linear models*, so we begin with the latter. In classical linear modeling, a sample of n independent observations $(y_1, y_2, ..., y_i, ..., y_n)$ is regarded as the realization of n independently distributed components $(Y_1, Y_2, ..., Y_i, ..., Y_n)$ of a random variable Y with means $(\mu_1, \mu_2, ..., \mu_i, ..., \mu_n)$. In the *systematic* part of the classical linear model, each mean, μ_i, is regarded as a linear function of p explanatory variables $(x_{i0}, x_{i1}, ..., x_{ij}, ..., x_{ip-1})$, usually with $x_{i0} = 1$ for the intercept, that is,

$$E(Y_i) = \mu_i = x_{i0}\beta_0 + x_{i1}\beta_1 + \ldots + x_{ij}\beta_j + \ldots + x_{ip-1}\beta_{p-1} = \textstyle\sum_j x_{ij}\beta_j \qquad (10.1)$$

where, for $i = 1, ..., n$, and $j = 0, 1, ..., p - 1$, x_{ij} is the value of explanatory variable j for observation i, and β_j is the parameter determining the direction and degree of association of μ_i with explanatory variable x_{ij}. Equation (10.1) is a *linear prediction function* or *linear predictor* whereby the expected values of the Y_i components $[E(Y_i) = \mu_i]$ are predicted by $\sum x_{ij}\beta_j$. The systematic part of the classical linear model assumes that all explanatory variables that influence the mean are included in the model and measured without error. In the random part of the classical linear model, each component Y_i is assumed to be independently distributed with a normal (Gaussian) probability distribution and constant variance $\sigma_i^2 = \sigma^2$ for all Y_i, $i = 1, 2, ..., n$.

In generalized linear modeling, the linear predictor is allowed to predict a chosen function of the mean, $g(\mu_i)$, the variance σ_i^2 is allowed to vary as a function of the mean μ_i, and the random variable Y_i is allowed to have any distribution in the *exponential dispersion family*, a large family including the normal, Poisson, binomial, negative binomial, and multinomial distributions. The generalized linear model thus subsumes and generalizes the classical linear model. Analogous to the classical linear model, the systematic component of the generalized linear model has the form:

$$\eta_i = g(\mu_i) = \Sigma_j x_{ij}\beta_j \tag{10.2}$$

where:

$$\eta_i = g(\mu_i) \tag{10.3}$$

is called the *link function*, because it links the mean μ_i to the explanatory variables, and:

$$\eta_i = \Sigma_j x_{ij}\beta_j \tag{10.4}$$

is the linear predictor in the special case of the classical linear model, the link function is the identity:

$$\eta_i = \mu_i \tag{10.5}$$

and, as shown in Equation (10.1), the linear predictor is:

$$\mu_i = \Sigma_j x_{ij}\beta_j \tag{10.6}$$

By contrast, in a *log-linear model* (for count data), the link function is:

$$\eta_i = \log \mu_i \tag{10.7}$$

so the linear predictor is:

$$\log \mu_i = \Sigma_j x_{ij}\beta_j \tag{10.8}$$

If, as in the example presented later in this chapter, a log-linear model is applied to rate data, where each count y_i is divided by an exposure measure v_i, then the link function is:

$$\eta_i = \log \mu_i / \nu_i = \log \mu_i - \log \nu_i \tag{10.9}$$

and the linear predictor is:

$$\log \mu_i / \nu_i = \Sigma_j x_{ij}\beta_j \tag{10.10}$$

or, equivalently:

$$\log \mu_i = \Sigma_j x_{ij}\beta_j + \log \nu_i \tag{10.11}$$

where the additive term log v_i is called an *offset*. In generalized linear modeling of rate data, the offset is modeled as an additional explanatory variable (covariate) in the model with parameter $\beta = 1$ forced to unity.

The random component of the generalized linear model specifies the response variable Y with independent observations ($y_1, y_2, ..., y_n$) drawn from a probability distribution in the exponential dispersion family, all members of which have the form:

$$f(y_i;\theta_i)=\exp\{[y_i\tilde{\theta}_i \quad \mathrm{b}(\theta_i)]/\mathrm{a}(\phi)+\mathrm{c}(y_i,\phi)\} \tag{10.12}$$

where θ_i is the *natural parameter*, ϕ is the *dispersion parameter*, and $\mathrm{a}(\phi)$, $\mathrm{b}(\theta_i)$, and $\mathrm{c}(y_i, \phi)$ are functions taking different forms for different members of the exponential family (for example, normal, Poisson, binomial, and so on).

The probability mass or density function for any member of the exponential dispersion family can be written in the form of Equation (10.12). In the case of the Poisson distribution, for example, we have:

$$f(y_i;\mu_i)=e^{-\mu_i}\mu_i^{y_i}/y_i!=\exp[y_i\log\mu_i-\mu_i-\log y_i!]=\exp[y_i\theta_i-\exp(\theta_i)-\log y_i!] \tag{10.13}$$

where $\theta_i = \log \mu_i$, $\mathrm{a}(\phi) = 1$, $\mathrm{b}(\theta_i) = \exp(\theta_i) = \mu_i$, and $\mathrm{c}(y_i, \phi) = -\log y_i!$. Although other link functions might also be considered, the link function for which $\mathrm{g}(\mu_i) = \theta_i$ is called the *canonical link*, whereby the natural parameter equals the linear predictor:

$$\theta_i = \textstyle\sum_j x_{ij}\beta_j \tag{10.14}$$

As can be seen in Equations (10.13) and (10.14), log μ_i is the canonical link function for a log-linear model and also usually makes the most sense in practical applications as it precludes negative predictions for count data.

Finally, the variance for any member of the exponential dispersion family is the product of the variance function $\mathrm{V}(\mu_i)$ and the dispersion function $\mathrm{a}(\phi)$, that is:

$$\mathrm{Var}(Y_i)=V(\mu_i)\ \mathrm{a}(\phi) \tag{10.15}$$

where the variance function:

$$V(\mu_i)=\mathrm{b}''(\theta_i) \tag{10.16}$$

is the second derivative of the function $\mathrm{b}(\theta_i)$ in Equation (10.12), and the dispersion function $\mathrm{a}(\phi)$ commonly has the form:

$$\mathrm{a}(\phi)=\phi/w_i \tag{10.17}$$

where the dispersion parameter ϕ is divided by a known prior weight w_i, which can vary for each observation Y_i, but is often unity, whereby the variance is $\text{Var}(Y_i) = \text{V}(\mu_i)\phi$. In the classical linear model with normal distribution and variance σ^2, the variance function is $\text{V}(\mu_i) = 1$, and the dispersion parameter is $\phi = \sigma^2$. In the Poisson log-linear model, the variance function is $\text{V}(\mu_i) = \mu_i$, and the dispersion parameter is $\phi = 1$.

10.1.2 Parameter Estimation and Statistical Inference

In *maximum likelihood estimation* of the generalized linear model parameters (β_0, β_1, ..., β_{p-1}), the likelihood of the sampled observations (y_1, y_2, ..., y_n) is expressed as a function of those parameters, and estimates of (β_0, β_1, ..., β_{p-1}) are found that maximize the likelihood, or rather the log likelihood, which yields the same estimates but is more mathematically tractable. The likelihood of an observation from the exponential dispersion family is:

$$f_1 = \exp\{[y_i\theta_i - \text{b}(\theta_i)]/\text{a}(\phi) + \text{c}(y_i, \phi)\} \tag{10.18}$$

and the log likelihood of an observation from the exponential dispersion family is thus:

$$L_i = \log f_i = [y_i\theta_i - b(\theta_i)]/a(\phi) + c(y_i, \phi) \tag{10.19}$$

The likelihood of a sample of n independent observations (y_1, y_2, ..., y_n) is the product of the n individual observation likelihoods f_i, that is:

$$f_1 f_2 \ldots f_n = \Pi_n f_i = \Pi_n \exp\{[y_i\theta_i - b(\theta_i)]/a(\phi) + c(y_i, \phi)\} \tag{10.20}$$

and the log likelihood of a joint sample of n independent observations (y_1, y_2, ..., y_n) is thus the sum of the n individual observation log likelihoods L_i, that is:

$$\begin{aligned}\log[f_1 f_2 \ldots f_n] &= \log f_1 + \log f_2 + \ldots + \log f_n \\ &= \Sigma_i L_i = \Sigma_i\{[y_i\theta_i - b(\theta_i)]/a(\phi) + c(y_i, \phi)\}\end{aligned} \tag{10.21}$$

In the case of the log-linear model, the focus of this chapter, the log likelihood for a sample of n independent observations is thus:

$$\Sigma_i L_i = \Sigma_i[y_i \log\mu_i - \mu_i - \log y_i!] \tag{10.22}$$

The latter term in Equation (10.21) expresses the sample log likelihood as a function of the known sample observations (y_1, y_2, ..., y_n) and unknown parameters (β_0, β_1, ..., β_{p-1}) that are estimated by maximizing $\Sigma_i L_i$ in maximum likelihood estimation. Estimates of β_0, β_1, ..., β_{p-1} are obtained as solutions of the likelihood equations:

$$\Sigma_i[(y_{\bullet i} - \mu_i)x_y / v(\mu_i)](\partial\mu_i / \partial\eta_i) = 0 \tag{10.23}$$

for $i = 1, ..., n$, and $j = 0, 1, ..., p - 1$, where η_i is the link function, and $v(\mu_i) = \text{var}(Y_i)$ is the variance (expressed as a function of the mean μ_i). The parameters β_j are implicit in Equation (10.23), since the mean is the inverse of the link function, that is, $\mu_i = g^{-1}(\Sigma_j x_{ij}\beta_j)$. For example, the variance $v(\mu_i)$ is σ^2 for a classical linear model, μ_i for a Poisson log-linear model, and $\phi\ \mu_i$ for a log-linear model which assumes that the variance is proportional to the mean to handle overdispersion. For the Poisson log-linear model, $\partial\mu_i/\partial\eta_i \mp v(\mu_i) = \mu_i$, so the likelihood equations simplify to

$$\Sigma_i[(y_{\cdot i} - \mu_i) \cdot x_y = 0 \quad (10.24)$$

for $i = 1, ..., n$, and $j = 0, 1, ..., p-1$. Because the likelihood equations are nonlinear functions of the parameters ($\beta_0, \beta_1, ..., \beta_{p-1}$) for most generalized linear models, maximum likelihood estimation requires iterative numerical methods. These methods, lucidly covered in McCullagh and Nelder (1989) and Agresti (2002) are beyond the scope of this chapter.

A generalized linear model with as many parameters as observations (that is, $p = n$) is called a *saturated model.* A saturated model perfectly fits the data, explaining 100 percent of the variance and yielding the highest possible maximum likelihood for the sample, but lacks scientific parsimony and other desirable properties such as a smooth curve fit. But the likelihood of the saturated model is a useful baseline for checking model fit. Let L_s denote the maximum log likelihood of a saturated model, and let L_a denote the maximum log likelihood of an alternative model with fewer parameters. For a Poisson or binomial model, the *scaled deviance* is twice the difference of the maximum log likelihoods of the saturated and alternative models, that is:

$$D^* = 2[L_s - L_a] = D/\phi \quad (10.25)$$

which expresses the deviance D as a multiple of the scale parameter. For Poisson or binomial models, $\phi = 1$, so $D^* = D$. Furthermore, for a Poisson or binomial modeling situation in which the number of sample observations n remains fixed regardless of counts, the deviance D divided by its degrees of freedom, $df = n - p$, has a chi-squared asymptotic distribution under the null hypothesis that the two models (saturated and alternative) fit the data equally well. Rejecting the null hypothesis indicates poor alternative model fit. On the other hand, failing to reject the null can indicate either good model fit or insufficient statistical power to detect a poor fit.

Analysis of deviance, a generalization of analysis of variance, is a powerful tool used to compare models and identify explanatory variables associated with variation in the criterion variable Y. Consider the comparison of two generalized linear models, M_0 and M_1, where M_0 is a special case of M_1. Then M_0 is said to be *nested* in M_1. For example, M_1 might include one parameter for a covariate not included in M_0, so M_0 would be the special case of M_1 with that parameter forced to equal 0; or M_1 might include an interaction term excluded from M_0; or M_1 might include a *set* of terms, such as interactions, excluded from M_0; and so on. Let D_0 denote the deviance of M_0, and let D_1 denote the deviance of M_1. Also let p_0 be the number of parameters in M_0, and let p_1 be the number of parameters in M_1 (with $p_1 > p_0$). The difference of deviances:

$$D_0 - D_1 = 2\log[\exp(L_1)/\exp(L_0)] \tag{10.26}$$

divided by its degrees of freedom, $df = p_1 - p_0$, is a *likelihood-ratio* statistic with a chi-squared asymptotic distribution under the null hypothesis that models M_0 and M_1 fit the data $(y_1, y_2, \ldots y_i, \ldots, y_n)$ equally well. Because a nested model M_0, with fewer parameters than M_1, can never fit better than M_1, $D_0 \geq D_1$, so a likelihood-ratio statistic is nonnegative. The larger the likelihood-ratio statistic, the worse the fit of M_0 compared to M_1, so rejecting the null hypothesis of no difference in fit indicates better fit of M_1 compared to M_0.

Maximum likelihood estimation also provides asymptotic (large sample) parameter estimate variances. The standard errors (variance square roots) of estimates of the model parameters β_j, $j = 0, 1, \ldots, p - 1$, may be of interest, for example, to construct confidence intervals. While the derivation of these estimates is beyond the scope of our discussion, McCullagh and Nelder (1989) and Agresti (2002) lucidly explain them.

In *maximum quasi-likelihood estimation*, an extension of generalized linear models, one need not assume a particular distribution for Y_i; instead, one assumes a mean-variance relationship:

$$\sigma_i^2 = \upsilon(\mu_i) \tag{10.27}$$

and substitutes the appropriate term for $\upsilon(\mu_i)$ in Equation (10.23). For the Poisson distribution, for example, where $\sigma_i^2 = \mu_i$, the quasi-likelihood equations substitute μ_i for $\upsilon(\mu_i)$ in Equation (10.23). The equations solved to obtain maximum quasi-likelihood estimates are exactly the same as the likelihood equations used in maximum likelihood estimation, but the equations are not true likelihood equations unless the Y_i distribution is a member of the *natural exponential family*, which is the subset of the exponential dispersion family where the dispersion parameter ϕ is known, for example, the Poisson distribution where $\sigma_i^2 = \mu_i$. Nevertheless, Wedderburn (1974) proposed quasi-likelihood estimation as a further generalization of generalized linear models to handle even more diverse situations and suggested using the estimating equations in Equation (10.23) with any variance function regardless of whether the underlying probability distribution belongs to the natural exponential family.

One important application of quasi-likelihood to analysis of log-linear model rate data is to handle overdispersion, that is, variances that exceed the means, as often occurs in practice with non-negative integer (count) data. Failure to correct for overdispersion increases the type I error rate (that is, true probability of erroneously rejecting the null hypothesis) for significance tests and erroneously reduces the width of confidence intervals. To handle this situation, the variance is assumed to be proportional to the mean, that is:

$$\sigma_i^2 = \phi \mu_i \tag{10.28}$$

where the dispersion or scale parameter ϕ is estimated and multiplied by the estimated mean to obtain the estimated variance corrected for overdispersion. The scale parameter ϕ can be estimated several ways, but a common method is based on the fact that the scaled deviance D/ϕ has a chi-square asymptotic distribution with $n - p$ degrees of freedom (and thus expectation $n - p$), so the deviance divided by the degrees of freedom is an estimate of the scale parameter, that is:

$$D/(n-p) \approx \phi \tag{10.29}$$

for large samples. Estimation for an overdispersed Poisson log-linear model proceeds by fitting the model by standard maximum likelihood methods, estimating the scale parameter ϕ using the full model deviance divided by its degrees of freedom, dividing log likelihoods used in likelihood ratio tests by the estimated ϕ, adjusting estimated parameter standard errors using the variance estimates multiplied by the estimated ϕ, and proceeding as usual with hypothesis tests and/or confidence intervals.

Finally, there are many well-known diagnostic procedures for testing the adequacy of a model, that is, plotting observed scores against predicted scores to identify potential outlier problems and plotting variances against means to assess the mean-variance assumption. See McCullagh and Nelder (1989) and Agresti (2002) for application of diagnostic methods in generalized linear modeling.

10.1.3 Illustrative Application

To evaluate the effectiveness of universal helmet laws, one approach is to compare motorcyclist fatalities in states with a universal helmet law to those in states without it, adjusting for differences in motorcycling activity between the states. Unfortunately, whereas the number of annual motorcycle registrations is available for individual states, the number of motorcycle miles traveled is not. Although the number of motorcycle registrations is related to exposure, it neglects variation in the activity of the registered motorcycles—a key quantitative measure needed to assess the association of fatality rates with helmet laws. Nevertheless, since motorcycling activity is highly seasonal, with more activity on warm or dry days than on cold or rainy days, and climates vary markedly across states in the United States, fatalities per registered motorcycle in the United States can be compared between states with and without universal helmet laws while controlling for climate measures correlated with motorcyclist activity.

This study employed maximum quasi-likelihood generalized linear modeling to explore the association of motorcyclist fatality rates with universal helmet laws using climate measures to control for motorcyclist activity (Morris, 2006). The analytic objective was to maintain scientific parsimony and statistical power, with minimal reliance on stringent statistical assumptions, by modeling fatality rates as a function of one explanatory variable (universal helmet law) and two climate-related activity measures (heating degree days, precipitation) along with pertinent quadratic and interaction terms. Quasi-likelihood generalized linear modeling provided crucial flexibility in modeling the relationship between a function of the mean and the covariates, the relationship between the mean and variance, and the error distribution.

Motorcyclist fatality data were from the National Highway Traffic Safety Administration's (NHTSA) Fatality Analysis and Reporting System (FARS) (NHTSA, 2005). FARS is a database of information about the scenarios, vehicles, drivers, and passengers involved in all fatal motor vehicle crashes on public highways and roads in the United States. Data on hospital emergency room-treated injuries were from the U.S. Consumer Product Safety Commission's (CPSC) National Electronic Injury Surveillance System All-Injury Program (NEISS-AIP) (CPSC, 2001). Data on the number of registered motorcycles by states were from the Federal Highway Administration (FHWA, 2005).

Normalized state climate data, including population-weighted annual heating degree days and precipitation inches, were from the National Oceanic and Atmospheric Administration (NOAA). The heating degree days statistic is a measure of cold weather energy consumption and is defined as the annual sum of daily differences in mean daily temperature from a 65° base (with the difference set to 0 if the mean daily temperature exceeds the 65° base temperature), averaged across all stations within the state, with the average weighted by population distribution in the area. At one station in a given year, for example, five days with a mean daily temperature of 64° would result in five degree days, as would one day with a mean daily temperature of 60°. NOAA's normalized heating degree day measure, an annual average derived over the 30-year period 1971–2000, is a climate measure that estimates the annual average heating degree days for each state during the normalization period. The advantage of heating degree days over average temperature as a measure of motorcyclist activity consists both in its theoretical utility for ratio-scale measurement of the change in thermal energy necessary to maintain a comfortable ambient temperature and in its empirical utility in accounting for substantial nuisance variation in fatality rates.

To demonstrate the seasonality of motorcyclist fatalities and injuries, and their strong association with climate measures, Table 10.1 gives monthly motorcyclist fatalities and injuries in the United States for 2001–2002 along with normalized heating degree days and precipitation inches for the coterminous United States. (The normalized climate measures by month were only available for the coterminous United States). Table 10.1 shows that the largest percentages of fatalities (11.1–13.5 percent) and injuries (10.7–13.1 percent) occurred during warm months (May–September) associated with the smallest percentages of normalized heating degree days (0.2–3.5 percent) and the largest percentages of precipitation inches (8.5–10.0 percent). Conversely, the smallest percentages of fatalities (2.6–3.6 percent) and injuries (3.2–3.7 percent) occurred during cold months (December–January) associated with the largest percentages of normalized heating degree days (16.2–20.3 percent) and the smallest percentages of precipitation inches (6.8–7.5 percent).

Table 10.2 confirms large statistically significant Pearson correlations among the monthly measures in Table 10.1, with a correlation of .98 for motorcyclist

Table 10.1 Monthly motorcyclist fatalities and U.S. hospital emergency room-treated injuries during 2001-2002 and normalized heating degree days and precipitation inches

	Motorcyclist fatalities, 2001–2002		U.S. Emergency room-treated motorcyclist injuries, 2001–2002		Normalized heating degree days (coterminous United States)		Normalized precipitation inches (coterminous United States)	
Month	Number	%	Number	%	Number	%	Number	%
1	170	2.6	8,098	3.6	917	20.3	2.27	7.5
2	222	3.4	8,370	3.7	732	16.2	2.04	6.8
3	332	5.2	11,652	5.2	593	13.1	2.59	8.6
4	549	8.5	21,868	9.7	345	7.6	2.44	8.1
5	713	11.1	23,938	10.7	159	3.5	3.01	10.0
6	850	13.2	28,476	12.7	39	0.9	2.92	9.7
7	842	13.1	25,888	11.5	9	0.2	2.79	9.2
8	870	13.5	29,364	13.1	15	0.3	2.65	8.8
9	796	12.4	25,348	11.3	77	1.7	2.58	8.5
10	504	7.8	19,296	8.6	282	6.2	2.29	7.6
11	359	5.6	14,900	6.6	539	11.9	2.40	7.9
12	234	3.6	7,188	3.2	817	18.1	2.23	7.4
Total	6,441	100.0	224,386	100.0	4,524	100.0	30.21	100.0

Sources: NHTSA, CPSC, NOAA, 2004.

Table 10.2 Correlation matrix for monthly motorcyclist fatalities and U.S. hospital emergency room-treated injuries during 2001–2002 and normalized heating degree days and precipitation inches

	U.S. emergency room-treated motorcyclist injuries	Normalized heating degree days	Normalized precipitation inches
Motorcyclist fatalities	0.983**	−0.983**	0.800*
U.S. emergency room-treated motorcyclist injuries		−0.979**	0.772*
Normalized heating degree days			−0.764*

*$p < .005$, **$p < .0001$; 2-tail.
Source: Bureau of Transportation Statistics, 2005.

fatalities and injuries, −.98 for fatalities and heating degree days, .80 for fatalities and precipitation, −.98 for injuries and heating degree days, .77 for injuries and precipitation, and −.76 for heating degree days and precipitation.

Figure 10.1 gives fatalities per 10,000 registered motorcycles per year as a function of universal helmet law and annual heating degree days for all 50 states. Fatality rates are linearly associated with annual heating degree days in both universal helmet law (R^2 = .284) and non-universal helmet law (R^2 = .519) states, with essentially parallel least-squares regression lines relating fatality rates to heating degree days. Range restriction in the universal helmet law states (n = 20) is the most likely explanation of the smaller proportion of variance in fatality rates accounted for by heating degree days in those states, which exclude both Alaska and Hawaii. Although there is dispersion about the regression lines in both groups, reflecting other relevant factors influencing state motorcyclist fatality rates, a substantial portion of that dispersion is attributable to variation in annual precipitation as shown in Figure 10.2.

Figure 10.2 gives fatalities per 10,000 registered vehicles per year as a function of universal helmet law and annual precipitation inches for all 50 states. Figure 10.2 reveals quadratic association of fatality rates with annual precipitation. The linear and quadratic components of precipitation inches together account for about 18 percent of the variance in fatality rates among universal helmet law states and 35 percent in non-universal helmet law states. It is beyond the scope of this analysis to attempt an explanation of why the relation of state fatality rates with state average annual precipitation should take the *J* form in Figure 10.2; rather, the purpose is to control nuisance variation in state fatality rates (that is, mainly due to variation in activity) to permit a parsimonious and statistically powerful assessment of the association of fatality rates with universal helmet laws. The full linear model relat-

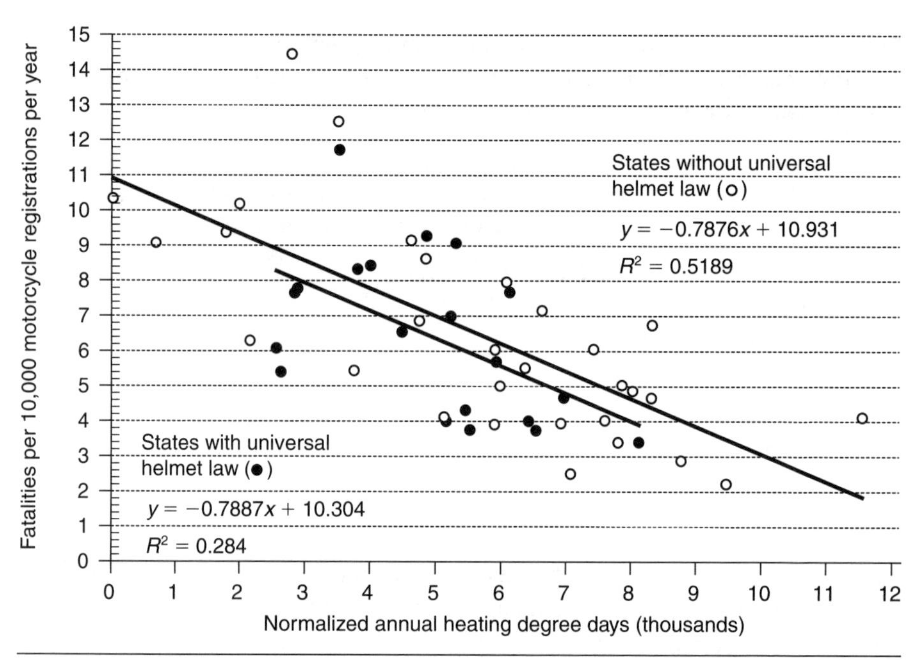

Figure 10.1 Motorcyclist fatalities per 10,000 registered motorcycles per year in states with or without a universal helmet law every year from 1993 through 2002 as a function of annual heating degree days. Source: NHTSA, FHWA, NOAA, 2004.

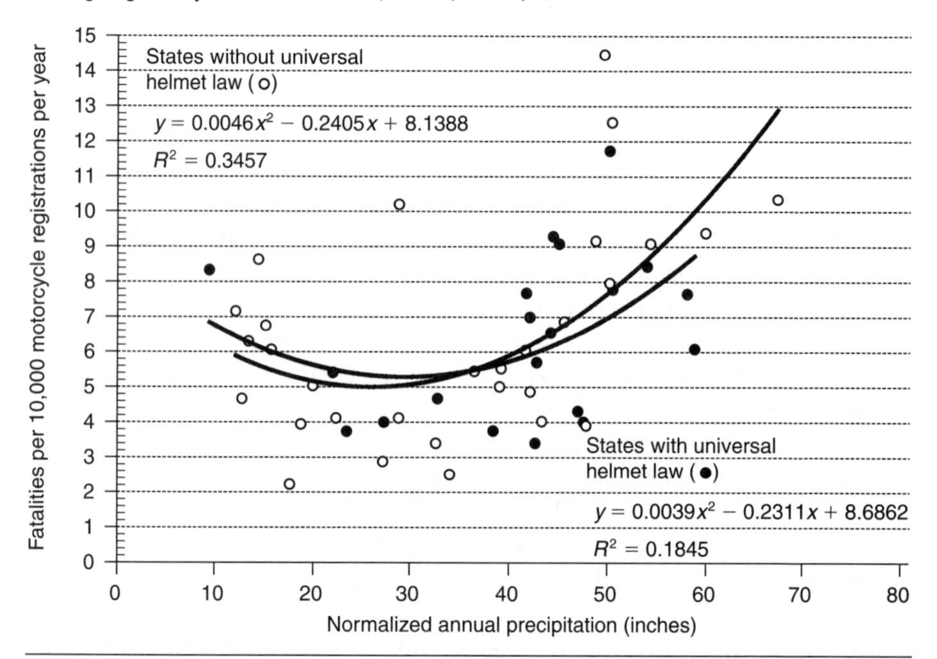

Figure 10.2 Motorcyclist fatalities per 10,000 registered motorcycles per year in states with or without a universal helmet law every year from 1993 through 2002 as a function of annual precipitation (inches). Source: NHTSA, FHWA, NOAA, 2004.

ing fatality rates to heating degree days, precipitation inches, squared precipitation inches (a quadratic term), the products of heating degree days with precipitation and squared precipitation inches (interaction terms), and the dichotomous universal helmet law indicator accounts for 55 percent of the variance in state fatality rates and 58 percent of the (natural) log-transformed fatality rates. Thus, normalized heating degree days and precipitation inches together account for substantial variation in state motorcyclist fatality rates.

Generalized linear regression analysis compared fatality rates in states with and without universal helmet laws adjusting for exposure as indexed by normalized state climate data. For each state, the annual average fatality rate was estimated by dividing the sum of fatalities across the decade from 1993 to 2002 by the sum of motorcycle registrations each year across the same period. The annual average heating degree days and precipitation were obtained likewise. To include all 50 states in the analysis, the five states that repealed a universal helmet law sometime during the 1993–2002 decade (3 in 1997, 1 in 1999, and 1 in 2000) were grouped with states that did not have a universal helmet law during that decade. Also, following NOAA, data for the District of Columbia were combined with those for Maryland, both of which had a universal helmet law throughout the 1993–2002 decade. The comparison is thus between 20 states in the United States that had a universal helmet law every year from 1993 through 2002 and 30 states that did not (that is, 25 states that did not have a universal helmet law and five states that did for some years, but not for the entire decade). Since helmet usage is known to correlate highly with universal helmet law requirements, this distinction is likely to correlate highly with actual helmet usage in the states, with higher total usage in states that had a universal helmet law in effect during the entire decade as compared to states that did not. An additional analysis was run in which the dichotomous universal helmet law grouping variable was replaced with a continuous variable measuring the number of years (rounded to the nearest 1/2 month) that a universal helmet law was actually in effect in each state from 1993 through 2002. The results of both analyses were similar.

To assess the association of fatality rates with helmet laws while controlling for annual heating degree days, precipitation inches, squared precipitation inches, and their interaction, quasi-likelihood generalized linear regression analyses were performed using the SAS (V8.0) GENMOD procedure and log-linear model:

$$\log(\mu_F / V) = \beta_0 + \beta_1 D + \beta_2 P + \beta_3 P^2 + \beta_4 DP + \beta_5 DP^2 + \beta_6 H \qquad (10.30)$$

where log $(\cdot)$ denotes the natural log function, μ_F = expected annual fatalities, V = annual motorcycle registrations (10,000s), D = annual heating degree days, P = annual precipitation (inches), and for comparison, either (a) $H = 0$ or 1 indicating whether the state had a universal helmet law every year from 1993 to 2002 or (b) $0 \leq H =$ years helmet law in effect from 1993 to 2002 ≤ 10.

Whether the universal helmet law factor was regarded as dichotomous or continuous, each analysis included an intercept parameter β_0 and parameters for linear association of log fatality rate with D and P (β_1, β_2), quadratic association with P (β_3), interaction of D with the linear (β_4) and quadratic (β_5) components of P, and association with a universal helmet law (β_6). Model parameters were estimated via maximum likelihood assuming a Poisson distribution, with parameter estimate variances adjusted for overdispersion via quasi-likelihood generalized linear modeling methods using the square root of the deviance divided by the degrees of freedom to estimate the generalized linear model scale parameter.

Analyses employed the SAS (V8.0) GENMOD procedure with the form:

```
proc genmod; model F = D P P2 DP DP2 H/
                 dist=poi link=log offset=LV scale=deviance type1 type3;
```

This SAS code specifies the model depicted in Equation (10.30), in which the variables are defined as previously stated, except that F denotes fatalities, $P2$ denotes P^2, $DP2$ denotes DP^2, and LV denotes the natural log of V. The modeling options specified after the slash (/) are defined as follows: dist = poi specifies the Poisson distribution; link = log specifies a log-linear model; offset = LV specifies the offset for this log-linear rate analysis; scale = deviance specifies that the scale parameter is to be estimated using the full model deviance divided by its degrees of freedom $n - p$; type 1 specifies a set of hierarchical analyses using the scale parameter estimated by the deviance obtained for the model including only the parameters in the model at that point (for example, after D and P are in the model); and type 3 specifies a simultaneous analysis using the scale parameter estimated by the deviance obtained for the full model depicted in Equation (10.30).

As shown in Table 10.3, all parameter estimates in the quasi-likelihood generalized linear model analysis differ significantly from zero, including the fatality rate reduction associated with the universal helmet law whether the latter was measured dichotomously [$F(1, 43) = 2.72$, $p = .053$, one-sided] or continuously [$F(1, 43) = 2.63$, $p = .055$, one-sided]. A one-sided test of the universal helmet law effect is justified by *a priori* expectation of a safety benefit from existing empirical and biophysical evidence. The overall fit of either generalized linear model with estimated scale parameter was excellent whether the universal helmet law was measured dichotomously [scaled $\chi^2(43) / 43 = 1.04$] or continuously [scaled $\chi^2(43) / 43 = 1.03$]. In conclusion, with climate measures statistically controlled, state universal helmet laws were associated with lower motorcyclist fatality rates. This finding is consistent with studies using a variety of methodologies that have also reported motorcycle helmet safety benefits (Norvell and Cummings, 2002; Sass and Zimmerman, 2000).

Table 10.3 Generalized linear regression results

	Universal helmet law factor					
	Dichotomous			Continuous		
Source	**Estimate**	**SE**	**F**	**Estimate**	**SE**	**F**
Intercept	−8.3587	0.6125		−8.4609	0.6056	
Heating degree days (D)	0.2783	0.1170	5.69*	0.2893	0.1163	6.20*
Precipitation (P)	0.0860	0.0364	5.97*	0.0939	0.0365	7.04*
P^2	−0.0011	0.0005	5.73*	−0.0012	0.0005	6.50*
DP	−0.0255	0.0071	13.27*	−0.0263	0.0072	13.97*
DP^2	0.0004	0.0001	14.99*	0.0004	0.0001	15.26*
Universal helmet law	−0.1284	0.0778	2.72**	−0.0145	0.0089	2.63**

*$p < .05$, **$p = .106$, ***$p = .112$; 2-sided.

Note: Results are for likelihood-ratio tests with a full model log-likelihood of 4569.0251 and a scale parameter of 5.5666 estimated by the square root of the full model deviance divided by the degrees of freedom (i.e., 5.5666 = $(\sqrt{1332.4549}/\sqrt{43})$ *and do not depend on the order of entry into the model.*

Source: Bureau of Transportation Statistics, 2005.

10.2 SUMMARY

The results show that climate measures have considerable promise as indirect measures (proxies) of motorcycling activity. And, more to the point of this chapter, illustrate the utility of quasi-likelihood generalized linear regression modeling in the analysis of fatality risk data.

The views in this chapter are those of the author and do not necessarily represent the views of the Bureau of Transportation Statistics, the Research and Innovative Technology Administration, the U.S. Department of Transportation, or any other agency or staff.

REFERENCES

Agresti, A. 2002. *Categorical Data Analysis*. Hoboken, New Jersey: Wiley.

Diggle, P., P. Heagerty, K.-Y. Liung, and S. Zeger. 2002. *Analysis of Longitudinal Data*. New York: Oxford.

Federal Highway Administration, U.S. Department of Transportation. 2004. Highway Statistics 1992–2002. Washington, DC. Annual vehicle-miles of travel and related data by highway category and vehicle type for each year.

from http://www.fhwa.dot.gov/policy/ohpi/hss/hsspubs.htm (accessed March 3, 2004).

McCullagh, P. and J. A. Nelder. 1989. *Generalized Linear Models.* New York: Chapman and Hall.

McCulloch, C. E. and S. R. Searle. 2001. *Generalized, Linear, and Mixed Models.* New York: Wiley.

Morris, C. C. 2006. Generalized linear regression analysis of association of universal helmet laws with motorcyclist fatality rates. *Accident Analysis and Prevention*, 38: 142–147.

National Highway Traffic Safety Administration, U.S. Department of Transportation. 2004. Fatality Analysis Reporting System (FARS). See http://www-nrd.nhtsa.dot.gov/departments/nrd-30/ncsa/FARS.html. December 30, 2004.

National Climatic Data Center, National Environmental Satellite, Data, and Information Service, National Oceanic and Atmospheric Administration. 2002. State, Regional, and National Monthly Heating Degree Days Weighted by Population (2000 Census) 1971–2000 (and previous normal periods), Historical Climatography Series No. 5–1. Asheville, NC. Downloaded from http://www.ncdc.noaa.gov/oa/climate/normals/usnormals.html#Release. December 30, 2004.

National Climatic Data Center, National Environmental Satellite, Data, and Information Service, National Oceanic and Atmospheric Administration. 2002. State, Regional, and National Seasonal Temperature and Precipitation Weighted by Area 1971–2000 (and previous normals periods), Historical Climatography Series No. 4-3. Asheville, NC. Downloaded from http://www.ncdc.noaa.gov/oa/climate/normals/usnormals.html#Release.

Norvell, D. C. and P. Cummings. 2002. Association of helmet use with death in motorcycle crashes: a matched-pair cohort study. *American Journal of Epidemiology*, 156(5): 483–487.

SAS Institute. 2005. SAS version 8.0, online documentation at http://v8doc.sas.com/sashtml/.

Sass, T. R. and P. R. Zimmerman. 2000. Motorcycle helmet laws and motorcyclist fatalities. *Journal of Regulatory Economics*, 18(3): 195–215.

U.S. Consumer Product Safety Commission. 2001. NEISS All Injury Program: Sample Design and Implementation. Washington, DC.

Wedderburn, R. W. M. 1974. Quasi-likelihood functions, generalized linear models, and the Gauss-Newton method. *Biometrika*, 61(3): 439–447.

11

DEVELOPING STATEWIDE WEEKEND TRAVEL-DEMAND FORECAST AND MODE-CHOICE MODELS FOR NEW JERSEY

Rongfang (Rachel) Liu
Department of Civil and Environmental Engineering
New Jersey Institute of Technology

Yi Deng
Parsons Transportation Group

11.1 INTRODUCTION

Realizing the benefits of statewide travel-demand models, more and more states have already been or are in the process of developing various statewide procedures, structures, or uniform models for travel-demand forecast and mode choices. This chapter intends to fill the gap by introducing an ongoing effort undertaken by the New Jersey Department of Transportation (NJDOT) to develop a statewide weekend travel-demand forecast and mode choice model in New Jersey. Applying a holistic approach that balances state-of-the-art research and practical modeling applications for multiple agencies, the research team has examined the unique characteristics of weekend travel and evaluated travel survey data and is developing specifications for a statewide weekend travel-demand forecast model that can be incorporated into the existing long-range transportation planning (LRTP) process at both metropolitan and state levels.

Many states have developed statewide travel-demand forecast models such as those in Horowitz (2006) that cover land use, intercity travel, toll facilities, freight movements, and other data. However, there are no statewide weekend travel-demand forecast models in the existing literature despite increased recognition of the rapidly growing pace of weekend travel.

The lack of attention to and limited information about weekend travel might result from its diversity of trip purposes, travel distances, and spatial and temporal distribution patterns. Interest might also be overshadowed by the overwhelming concern about congestion during weekday peak periods. However, recent studies and our travel experiences prove that congestion is occurring more and more frequently on weekends; furthermore, because days on which emissions exceed allowable levels often occur on summer weekends, air quality concerns have drawn the attention of many planners and decision makers.

Person-trip rates during weekends are only marginally lower than those during weekdays. Thus, despite the large amount of staff time and funding spent on analyzing weekday commuting travel, it is essential that we draw attention to weekend travel. As demonstrated in Table 11.1, the survey data collected in the North Jersey Transportation Planning Authority (NJTPA) jurisdiction show a difference of only 0.4 to 1.4 percent from two sources (PBQD, 2000).

As one of the most densely populated and transit-friendly states, New Jersey has a substantial share of transit modes. Given the large numbers of trips for purposes other than commuting during peak hours, it is not difficult to conclude that there is a risk that the need for additional highway and transit capacity is sig-

Table 11.1 Weekday and weekend trip rates

	RT-HIS		NPTA	
	Weekday	**Weekend**	**Weekday**	**Weekend**
Sample size (number of households)	4,541	275	321	128
Estimated mean (number of trips per HH)	8.80	7.71	10.35	8.53
Difference between weekend and weekday	0.4%		1.4%	

nificantly understated when relying on existing models. The result of such understatement could place New Jersey at a disadvantage in securing federal funding for capital projects. It is therefore essential to develop new models that more accurately predict multimodal travel for non-work purposes, especially on weekends.

11.2 RESEARCH PROBLEM AND BACKGROUND

Current travel demand forecast models, instigated by the federal mandate (U.S. Congress, 1991) of the LRTP for metropolitan areas, are reasonably accurate in projecting work trips and peak-hour travel. However, national trends reveal that work trips are becoming a declining portion of total trip-making, and off-peak travel volumes on weekends can often exceed peak-hour congestions.

According to the National Household Travel Survey (NHTS, 2001), as shown in Figure 11.1, the share of commuting trips in terms of vehicle miles traveled (VMT) has decreased from 34 percent in 1969 to 27 percent in 2001. The same trend was confirmed by intercity travel as documented in the American Travel Survey (ATS, 2001).

The characteristics and dynamics of weekday and weekend travel are likely different even if the magnitude of travel per household is comparable. According to Lockwood, Srinivasan, and Bhat (2005), the weekend activity-travel participation is largely nonwork-oriented, while weekday activity participation is centered on work or other mandatory activities. Weekend travel is not likely to follow the same peaking characteristics as weekday travel. Special traffic generators such as sports activities, concerts, and other cultural events during weekends result in traffic characteristics that are dissimilar to the typical weekday traffic.

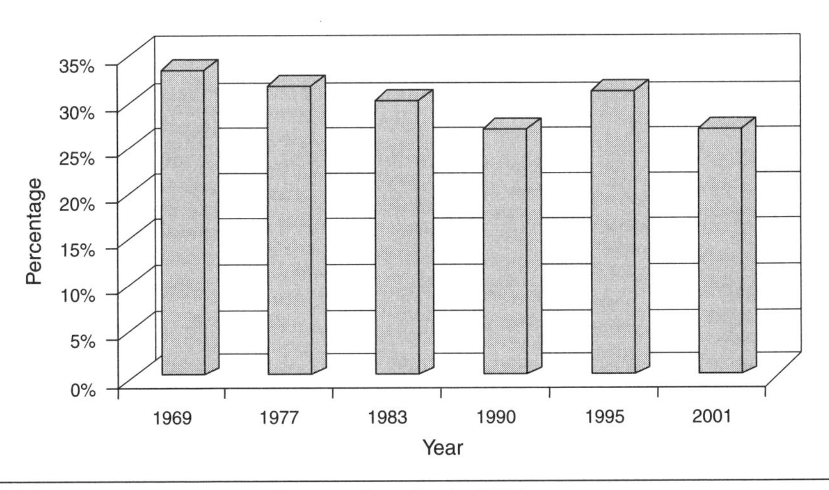

Figure 11.1 Decreasing shares of commuting trips by VMT

In addition, air pollution concerns, demand shift from weekday to weekend, and general social consideration of promoting physically active recreational pursuits, strengthen the argument that weekend travel warrants careful attention for comprehensive travel-demand modeling as well as for evaluating policy actions aimed at alleviating congestion, improving air quality, and enhancing the overall quality of life.

The purpose of this research is to specify an ultimate model to forecast weekend travel that incorporates the following processes: trip generation by trip types, time of day, origin-destination pattern, and mode choice. To derive the ultimate product of this effort, the research team has conducted a series of interviews, surveys, and model specifications for the statewide weekend travel-demand forecast model in New Jersey. The following sections present highlights of the research.

11.3 CURRENT STATE OF PRACTICES IN WEEKEND MODELS

The research team conducted two broad surveys to determine the current state of practice in weekend travel demand forecasting. The first survey was conducted using the Travel Model Improvement Program (TMIP) list-server, an Internet discussion group of modelers, including academics, consultants, and government employees. The second survey was distributed among selected Metropolitan Planning Organizations (MPO) in North America to determine the current status of weekend travel demand and forecast models.

11.3.1 TMIP Discussions

The TMIP was established about 10 years ago by the Federal Highway Administration (FHWA) to help planning agencies improve techniques for informing decision makers about how growth in population and employment, development patterns, and investments in transportation infrastructure are likely to affect travel, congestion, air quality, and quality of life. To advance the state of the practice of travel modeling and planning analysis, TMIP provides a variety of services to academics and professionals, ranging from seminars and training to e-mail lists, clearinghouses, research, and peer review and exchanges (USDOT, 2006).

In April 2006, Dr. Liu, the principal investigator of the project, posted an invitation to discuss weekend travel behavior and forecast models. The e-mail asked about three aspects of weekend travel demand modeling: current weekend modeling development, factors that impact weekend travel, and data collection on weekend travel characteristics.

The e-mail invitation has stimulated interesting discussions among modelers from MPOs, academia, and consultants. The microsimulation model of weekend travel by households in Calgary, Alberta, Canada, had been identified as in the most advanced stage of MPO modeling and is briefly described in this section. A series of travel-behavior and transit onboard surveys that include a weekend element has also been identified but is not included.

11.3.1.1 Differences Between Weekday and Weekend Travel

One of the most important contributions from this list-server discussion is its insights into the unique characteristics of weekend travel and its spatial and temporal distribution in various geographic locations. As pointed out by one of the responses (Leve, 2006), there are several cities in which significant numbers of people leave the urban area on weekends, typically for various activities associated with time in the country. The traffic patterns associated with large numbers of people leaving on Friday evening, and to a lesser extent early on Saturday morning, and returning on Sunday evening are quite different from typical weekday traffic patterns. Nevertheless, these traveling characteristics can cause significant and extensive congestion. In addition, the congestion can be more perceptible in outlying areas that do not normally have congestion problems.

The weekend tourist travel behavior is certainly more common during the summer months when atmospheric conditions can amplify the effects of local ozone concentrations. In modeling this type of trip, it is important to be aware that a reasonably well-defined area outside the city attracts many trips: the challenge is to predict who might be going to this area and from where.

11.3.1.2 Importance of Analyzing Weekend Travel

In addition to identifying various sources of modeling development and data availability, this list-server discussion produced another important result, the reassurance of weekend travel-demand forecast and mode-split analysis, as well as its overall impact on the LRTP process.

The characteristics of weekend and weekday travel are quite different, as pointed out by various responses in the list-server discussion. Unique traffic generators such as sporting events and concerts during some weekends could result in traffic congestion at different network links than those that are congested during the typical workday traffic profile.

The differences between weekdays and weekend days, especially in the temporal profiles of the travel patterns, can have implications for air-quality modeling. Specifically, the sustained high volume of weekend trips during the hotter midday period can amplify the severity of the impact of emissions on air quality. Further, as a consequence of departure from home much later in the day than on weekdays, the longer soak times of vehicles prior to first use during weekends could also increase air pollution from emissions. As pointed out by Dr. Bhat, of the three days exceeding the 125 parts per billion (ppb) ozone level nonattainment standard in the Dallas–Fort Worth area in 2003, two were weekend days, according to a North Central Texas Council of Governments (NCTCOG) report.

Furthermore, when looking at permanent vehicle recorder stations on interstate highways in many portions of the state of Washington, Shull (2006) stated that the highest travel hours are often on the weekends. This does reflect a high degree of intercity travel, but it also shows that we must pay more attention to that than typical weekdays. As we extract the last bit of capacity from our systems by using former shoulders as lanes, and so on, we need to keep in mind that the incidents, special cases, holidays, disasters, and other situations continue to require more and more of our focus (Shull, 2006). It is also possible that transportation control measures intended to regulate traffic flows on weekdays might transfer traffic to weekends.

11.3.1.3 Scope of Weekend Travel Demand Model

The overall response from the TMIP discussion acknowledges the need and supports the purpose of the weekend travel-demand forecast and mode-split model but cautions that care should be exercised on the scope of and investment in such modeling. As suggested by Cervenka (2006), a clear purpose needs to be stated before substantial investments are made in the development of weekend-based models. Perhaps rather than full regional weekend models, special event and subarea types of models are needed in which the survey/data collection program can be focused on well-stated objectives.

Since concerns about air quality are typically associated with the summer season, it is also suggested that one of the bigger deficiencies in current regional modeling activities is that they are often based on weekdays while schools are in session rather than on summer traffic. We certainly need to acknowledge some of the strong and philosophically correct arguments that are made from time to time in support of weekend modeling and weekend-based problem resolution. For example, even minor increases in ozone emissions during summer weekends have a measurable impact on a region's health simply because there are more people spending time outdoors on weekends. If one is to develop a weekend model, perhaps it needs to be specifically a summer-time weekend model.

11.3.2 MPO Survey

As stated earlier, most existing transit experts, MPOs, and statewide travel-demand models explicitly model nonwork travel purposes and include both peak and off-peak periods, but only for a typical weekday. However, systematic or general methodologies for estimating travel demands, and mode choice in particular, do not exist for weekend travel analysis, which we know is dominated by nonwork travel as well as specific peak and off-peak periods. It is our understanding that improvements in the analysis and forecasting of weekend travel and transportation impacts are the primary focus of this research project.

An in-depth MPO survey was conducted to gather weekend travel-demand analysis based on two major focuses. The first focus deals with the current status of weekend travel-demand model, data collection, and travel behavior, particularly on mode shares. The second focus is on the future plan of each MPO: do they plan to develop a weekend model, are there any factors they think are important in forecasting weekend travel, or are there modeling structures they would like to suggest. The research team selected the top 45 MPOs in terms of population, of which 20 responded. The following section summarizes the results based on the two focuses.

11.3.2.1 Current Landscape of Weekend Travel by MPOs

None of the MPOs surveyed has a weekend travel-demand forecast module in its LRTP model. However, Southern California Associations of Government (SCAG) is planning to issue a request for proposal (RFP) for developing a weekend travel-demand forecast. The weekend forecast model will be parallel to its original weekday model and based on a four-step forecast structure. The driving force behind this initiative is clearly a concern for air quality in various corridors. Many of the days that exceed air emission standards in the southern California region are weekend days.

Among the 20 responses to the survey, four have completed individual household travel behavior surveys since 1995. Among these household surveys, weekend travel information has been collected by SCAG, the Metropolitan Transportation Commission (MTC), the Atlanta Regional Commission (ARC), and the Oregon Department of Transportation and Service of Metropolitan District. Most of the data set and summary reports of the household surveys are accessible from the Internet, except the data set from SCAG.

The survey asked MPO staffs to compare weekend travel and weekday travel in their region. Most survey respondents emphasized that the traffic descriptions of weekends are based on their own experiences rather than collected data or analyses. MPO staffs observed different travel patterns on weekends and weekdays, even without solid empirical data. For example, one of the MPOs in Illinois reported that shopping and other major destinations attract higher volumes/ridership on weekends, depending on the facility and time of day. One of the MPOs in the northwest region found that weekday and weekend congestions occur in different locations; some of the facilities are more heavily congested over weekends. Another MPO in the northeast observed different functions of highways, such as commute-oriented highways and more vocational-oriented highways. For the commute-oriented highways, volumes are lower on weekends. However, depending on the time of the year, the more vocational highways, such as I-495, can expect more volume on weekends; this highway is used by many New York and New Jersey people traveling to Maine in the summer.

11.3.2.2 Future Plans for Weekend Travel Demand Forecast

However, most agencies, except SCAG, which is preparing to initiate an RFP, do not have a plan to incorporate weekend travel into their current travel-demand forecast mode/split models in the near future. They do not have a clear plan for developing weekend models either. But further probing indicates that more and more agencies, especially those in large metropolitan areas, are confronted by various congestion problems that occur in nontraditional, outside of peak, commuting periods. Some of the agencies, such as the Houston-Galveston Area Council, have developed factors to reflect the air quality conditions. Others, such as the Maricopa Association of Governments, are contemplating the option of incorporating weekend travels by capturing recreational behaviors.

On the other hand, most MPOs have put the weekend travel demand forecast models on the back burner. For example, a staff member from SCAG mentioned that they might consider it after this round of RFP processes. Another staff member from the Sacramento Area Council of Governments mentioned that they are currently developing an activity-based model and hoping that it will be a better base for weekend travel-demand forecasting.

The consensus is clear that the first step toward understanding people's behavior over the two weekend days is the development of a household travel behavior survey to capture the entire period from Friday afternoon to early Monday morning. However, as demonstrated by one New Jersey respondent, a local household travel behavior survey may not provide enough data because recreational attractions might also attract people from out of the state, therefore, he suggested that an external cordon survey needs to be included.

The suggestions for the modeling structure ranged from applying a simple factor to complex four-step models, to tour- or activity-based models. The majority of responses are along the line of traditional four-step models with an emphasis on trip generation and mode split steps. A number of responses identified activity- or tour-based models and also recognized the increased cost and effort entailed developing such models. Is there a consensus? One respondent stated that the benefit is largely-based on the questions that will be answered. This is a valid statement. For example, weekend congestion around regional shopping centers or tourist destinations could be analyzed by assigning estimated trip tables from traffic counts. Analysis of weekend regional air quality concerns could require traditional four-step models.

11.3.2.3 Roles of Special Generators

As directed by the project client, New Jersey Transit (NJ Transit), the survey team has added a question on special generators to the MPOs that we have surveyed in the later stages. As expected, a number of land-use types have been modeled by various travel-demand forecast models, among them airports, medical centers, colleges, ports, stadiums, retail malls, science centers, and downtown centers. One of the MPOs in Texas included non-residential adjustment by factoring hotel room occupancy rates. Another MPO in Arizona estimated the trip generation rate for airports and universities by using gross factors. The Puget Sound Regional Council actually modeled three major exhibition locations not for sporting functions such as football and baseball, but rather for their use for exhibition purposes. The gross factors for trip generation based on the exhibition function are derived from regional population bases. The same MPO also modeled ports for heavy truck traffic because over the years port volume has increased twice as fast as the population.

11.4 EXISTING TRAVEL DEMAND FORECAST MODELS IN NEW JERSEY

In New Jersey there are at least six existing travel-demand models focused on or related to the Garden State, as shown in Table 11.2. A brief review of each model is provided, which serves as the basis for developing a weekend travel-demand forecast and mode-choice model for New Jersey. Due to diversified travel patterns, modal preferences, and capabilities of each jurisdiction, the models in New Jersey vary by size, complexity, and structure. The following sections describe each travel-demand model pertaining to the weekend travel-demand forecast model development.

It is also important to assess the need and ability to develop and maintain a weekend travel-demand forecast model for each agency and to understand how the proposed model interacts with the existing weekday travel-demand forecast model. The project team has conducted interviews with the MPO's staff to identify expectations and preferences for a weekend travel-demand forecast model by each agency. The interviews are also recorded in the following sections.

Table 11.2 Existing travel demand forecast models in New Jersey

Agency	Platform	Weekend module	Weekend data
NJTPA	TRANPLAN	No	Small amount
SJTPO	Cube/TP+	Yes	Some
DVRPC	TRANPLAN w/Evans Algorithm	No	Small amount
NYMTC	TRANSCAD	No	Small amount
NJ Transit	TP+	No	Some
NJ DOT	Cube	No	None

11.4.1 Model Structures of Various MPOs

The NJTPA, along with NJDOT, maintains the North Jersey Regional Transportation Model (NJRTM), which covers the thirteen northern counties of New Jersey. This is a traditional four-step model that focuses on weekday commuter patterns. It has a transit component module that takes into account intra-New Jersey transit trips as well as trans-Hudson trips to and from New York City, but the component has not been updated for several years. One could say that this model is more highway-based. It can forecast traffic for the morning and evening peak periods and the off-peak periods.

Similarly, NJ Transit maintains a version of the northern New Jersey model that is based on the NJRTM but has replaced the mode-choice and transit models with analysis tools that are focused on better transit forecasting for New Jersey and trans-Hudson trips to and from New York. There is currently a model being developed to combine the NJTPA and NJ Transit models into one integrated model.

The Delaware Valley Regional Planning Commission (DVRPC) maintains a travel-demand model for its region that covers four central New Jersey counties. It is a traditional four-step model containing both highway and transit components. The current working version of the model forecasts daily traffic only, but there are versions that model generic peak and off-peak periods. No peak-hour model *per se* is available at this time. The DVRPC model is a weekday commuter-based model.

The South Jersey Transportation Planning Organization (SJTPO) maintains a travel-demand model for four southern counties in New Jersey. This model is an enhanced four-step model. It contains 24 trip purposes and is the only model in New Jersey that attempts to explicitly model special generators and recreational travel. It was originally developed from a beach survey done in 1995 but is primarily calibrated to a summer Friday evening peak. It has the coding in place to generate weekend/recreational-based trip tables by pivoting off the Friday trip table, but it has not been calibrated to weekend travel.

The Statewide Truck Model is maintained by NJDOT and is used primarily as a tool for analyzing on-road goods movement in the New Jersey region of the northeast corridor. The model contains all of the primary roadways in New Jersey plus portions of Delaware, Pennsylvania, the New York City boroughs, and other bordering downstate New York counties. Truck travel was initially developed using commodity flow data and has been enhanced with data from regional estimates of households and employment broken down into various Standard Industry Code (SIC) categories. Trip-generation equations were derived and compared to models developed in Phoenix (1991), San Francisco (1993), and Washington, DC. Automobile travel is derived in a trip table merging a process involving the three MPO models. As such, this model is also a weekday commuter-based model. The Statewide Truck Model was initially a daily assignment model only but has been updated recently to peak period/hour models for the use in the Portways and CPIP projects.

As one of the major employment and activity centers for New Jersey, New York City attracts a large proportion of residents from New Jersey and is the integral part of the New Jersey travel-demand forecast model. The New York Best Practice Model (NYBPM) is unique to the New York City region and might not have direct applicability to New Jersey MPO models. However, the techniques developed and travel behavior observed in the modeling development process can be conducive to the further research of weekend travel and more universal applications.

The NYBPM model consists of four consecutive modules:

- Household synthesis, auto-ownership, and journey frequency (HAJ)
- Mode and destination and stop choice (MDSC)
- Time of day choice (TOD)
- Highway and transit assignment

It predicts the total number of households by income, size, number of children, number of workers, and number of autos, and then determines the number of journeys that are produced for each subgroup over a 24-hour period. The MDSC module replaces the traditional trip-distribution and mode-choice module. Based on the person and household characteristics and land-use densities around the journey origin, this module predicts which modes of travel each person chooses, where the person goes, and if the person stops along the way. The current version of the NYBPM has a simplified timing model based on a set of predetermined look-up tables (often referred as peak/off-peak factors) with percentages of journeys by time periods. The look-up tables are stratified by journey purpose, leg, mode, and some aggregate spatial categories.

11.4.2 Expectations of Statewide Weekend Model

The research team has conducted interviews with all MPOs in New Jersey and the New York Metropolitan Transportation Council (NYMTC) in addition to reviewing the modeling structure and documents of each organization. After general discussions on their current travel-demand forecast model and potential integration with the weekend model we are developing, MPO staffs were invited to provide their expectations, which are documented below.

Focusing on the travel conditions in the jurisdictions of the NJTPA, the staffs mentioned several important aspects to be addressed in the weekend travel study. For example, the shopping trips on Saturdays in Bergen County are extremely high because all the stores are closed on Sundays. There is a large volume of trips from North Jersey and Manhattan to the northern New Jersey shore area, such as Ocean and Monmouth counties, which were not addressed in the NJRTM or the SJRTM. The majority of trips to the Pocono Mountains occur on weekends. An origin-destination (O-D) survey at bridges both with and without tolls over the Delaware and Hudson rivers will help identify the trips that originate from North Jersey, Manhattan, and other areas of New Jersey.

As for the future directions of weekend travel-demand forecasting and mode-choice models, NJTPA staffs expressed their strong desire to be informed and involved in the development process. Given the limited knowledge of weekend travel and the lack of research or development of weekend models, a consensus has been reached that the next immediate step is to collect data. When comparing the options of a focused effort of weekend data collection and a general household survey, including weekend data, NJTPA staffs tended to suggest that a weekend travel-data collection might be needed because large surveys have already been accumulated for the weekday commuting trips. NJTPA staffs also suggested that the project team should collaborate with other ongoing data collection efforts, such as the Newark bus service survey in North Jersey.

As for the expectations for the weekend travel-demand model, NJTPA staffs expressed their desire to have a model that can be integrated with their existing weekday models. The weekend model should be able to identify the hot spots in various places, especially if those hot spots are different from weekday hot spots or show worse congestion than weekday places. It is also important to identify and correlate the potential for some discretionary trips which shift between weekday and weekend due to excess weekday congestion. The staffs also suggested that a number of sub-area/market-based models rather than one statewide model might deliver better or at least more-targeted results.

Compared to the anticipation and high hopes of the weekend model by NJTPA staff, the reactions of the DVRPC staffs are quite different. As we were told, a weekend travel-demand model is not on the priority lists of the DVRPC

LRTP process. Their focus is still the weekday commuting trips because there are no focused weekend trip attractors as predominant in the South Jersey areas. Dr. Thomas Walker is willing to provide a zonal structure and Social Economic (SE) data from the DVRPC if we are going to develop weekend models. He is also willing to review the modeling structure of a proposed weekend travel-demand forecast model for New Jersey.

On the other hand, the expectations of the NYMTC staffs may be falling in between the NJTPA and DVRPC staffs. The research team interviewed the NYMTC staffs on March 9, 2007. All sections of the NYMTC Department of Technical Services, Model Development, Model Applications, and Travel Survey were represented. All attendees at the interview agreed that weekend travel is an important segment of overall travel in the tri-state area, but that information is lacking. Greater attention is needed for weekend travel and more data should be collected on weekend travel to understand its substitute and complimentary patterns to weekday travel and to improve the capabilities of MPOs in forecasting weekend travel.

As documented, there are a number of different travel-demand forecast models in New Jersey and surrounding areas. These models are maintained by various MPOs, DOT, and transit agencies, and were built on different platforms, ranging from TRANPLAN, TP+, and TRANSCAD. After a thorough review of the existing travel-demand models in New Jersey and interviews with the MPO staff, the research team obtained a detailed assessment of the modeling processes employed as background for how each of these models can be used or adapted in the weekend forecasting process in the tasks to follow.

11.5 SPECIFICATIONS FOR A STATEWIDE WEEKEND TRAVEL-DEMAND MODEL

As discussed earlier, the conventional four-step travel-demand forecast models adopted by various agencies in New Jersey are widely used and accepted to forecast both automobile and transit travel demand for a typical weekday, generally a nonsummer weekday when work and school activities are routine. The focus is also primarily on peak period travel, in which regular work and school travel dominate. Given the much larger share of nonmandatory or discretionary travel that occurs during the weekends, the greater influence of seasonality on these activities, along with the substantial scheduling and impacts of major weekend special events, it is not surprising that the current models are not an adequate platform to meet the challenges of modeling weekend travel demand or to forecast it for either road facilities or transit services.

There is a large spectrum of variables, including but not limited to demographic and social factors, economics, transportation, and land use that affect non-work travel. The directions and magnitudes of impact on non-work travel also vary according to different studies in different areas. Given the wide spectrum of existing modeling techniques, ranging from simple, direct, low-cost spreadsheet models to extensive activity-, trip-, or journey-based models, it is too early to eliminate any form or to be biased toward any individual specifications. The research team is in the process of exploring various models such as:

- A geographic model for bus service planning and marketing
- A route-based approach to estimating bus ridership
- A multivariate time-series model
- A multiple linear regression analysis of factors contributing to transit use

The research team considers a range of alternative methodologies as part of a proposed set of model specifications for possible weekend travel-demand models to be developed in this research initiative. These will likely include a mix of sketch-planning spreadsheet supported tools, along with network-based methods and demand estimation procedures similar to the conventional weekday models.

Even if the conventional four-step modeling approach is used as the basic framework for weekend travel forecasting, each of the models will undergo a significant transformation to reflect the following specific aspects of weekend travel behavior:

- Household trip production will include more detailed segments of discretionary travel (for example, by singling out planned joint activities, visiting relatives and friends, all-day travel, and so on).
- Zonal trip attraction models will require more detailed segmentation by land-use type with possible singling out of unique facilities (for example, major sport arenas).
- Trip distribution models will reflect generally longer distances observed for weekend discretionary travel as well as the known phenomenon of positive utility of travel when travel time/distance are intentionally kept at a certain level and are not minimized within certain limits.
- Mode-choice models will reflect a different set of preferences compared to regular weekday models that include a significantly higher share of joint travel, greater sensitivity to transit service convenience, greater share of taxi mode and generally higher willingness to pay (so-called situational Value of Time (VOT)).
- Peak factor/time-of-day choice model will have a completely different structure with, perhaps, additional segmentation by Saturday/Sunday and summer/winter behavior.

Given the strong diversity of activities that determine weekend travel patterns and their varied impacts on the capacity of transportation infrastructure and services, we anticipate that different approaches may be needed to address relevant issues for specific facilities, which serve specific corridors and specific travel markets. This approach is aimed at developing the most efficient and practical methods needed to address what we understand are at least some of the current planning issues that have been identified by NJ Transit as important, and require better weekend analysis and forecasting:

- Rail and bus travel to New York
- Intra-Jersey bus travel on certain routes
- Bus travel to Cape May and other New Jersey coast locations
- Hudson-Bergen LRT

The research team is in the process of integrating the weekend mode-choice models within the existing NJ Transit NJTDFM framework. In this task, the project team will work with NJ Transit to develop a model framework that attempts to balance the need to forecast accurately weekend transit patronage with realistic data-management requirements.

11.5.1 Development of Weekend Total Demand for Transit Travel and Trans-Hudson Automobile Travel

As stated previously, NJ Transit NJTDFM is a modified four-step travel-demand forecasting model in which Steps 1 and 2 (trip generation and distribution) are replaced by survey-derived trip tables (transit trips and trans-Hudson automobile trips) and NJTPA automobile trips (intra-New Jersey automobile travel). AECOM Consult was responsible for developing the underlying survey database for the NJTDFM. Some of these surveys, notably spring 2001 PATH and the 1996 trans-Hudson automobile survey, obtained data for a Saturday and a Sunday and may be used directly in the process. Others, including rail, bus, and ferry surveys, were obtained for average weekday periods only. Total demand for intra-New Jersey automobile travel will also need to be obtained from data from NJ Transit, NJTPA, or other external sources (possibly NYMTC).

The research team is working with NJ Transit to identify any existing surveys or data that would assist in understanding:

- Existing weekend commuter rail, bus, LRT, and ferry travel patterns (particularly to and from Manhattan)
- Weekend trip purposes (work-related travel versus discretionary travel)
- Other drivers of nontraditional weekend transit travel (sporting events, concerts, theater, and other regional cultural events)

11.5.2 Development of Weekend Transit Level-of-Service Matrices

Weekend transit level-of-service (skim) matrices are developed to accurately forecast weekend travel demand by mode. The weekend level-of-service matrices will serve to characterize the weekend transit services available throughout the region. These transit skims, along with the automobile level-of-service characteristics (from MPO model), will be used as the key input into the weekend mode-choice models. The research team is working with NJ Transit and NJDOT to identify methodologies that will, in a cost-effective manner, accurately represent weekend transit service in the region. Potential approaches for the development of weekend transit level-of-service matrices include:

- Developing and maintaining a new weekend transit network for a network-based approach
- Borrowing weekday off-peak networks as surrogates for the weekend transit networks
- Developing an alternative coding approach to facilitate the development of a spreadsheet type model for weekend travel

A conceptual standard structure for New Jersey will be developed based on the existing modeling applications and future improvement toward the ultimate long-term application.

11.6 SUMMARY

The research team is in the process of structuring the model calibration and validation to include careful comparison of the model results to the available sources of information with a possible subsequent adjustment of the modal parameters to better match the established targets. Validation targets will include the following typical sources of information:

- Expanded aggregate trip generation, distribution, and mode-choice statistics from the household surveys
- Weekend traffic and transit counts
- Transit statistics from the on-board transit survey
- Independent statistics on attendance and capacity of special events by location

Additional available sources of information from various agencies, as well as possible focused surveys, will be considered. The corresponding survey techniques will be developed based on the experience of the other regions and research works reported in the literature.

In this step, the research team will calibrate a proposed model structure from the research elements of the assignment. If the next stage of the project is funded, the research team will develop calibration targets from existing NJ Transit survey data. If a network-based solution is chosen, the research team will implement the final calibrated mode-choice model within a suite of mode-choice models. It is also important for the research team to validate that the weekend model replicates observed transit travel behavior.

REFERENCES

Bhat, C. R. and R. Gossen. 2004. A mixed multinomial logit model analysis of weekend recreational episode type choice. *Transportation Research Part B*, 38: 767–787.

Bhat, C. R. and A. Lockwood. 2004. On distinguishing between physically active and physically passive episodes and between travel and activity episodes: An analysis of weekend recreational participation in the San Francisco Bay area. *Transportation Research Part A*, 38: 573–592.

Bhat, C. R. and S. K. Singh. 2000. A comprehensive daily activity travel generation model system for workers. *Transportation Research Part A*, 34(1): 1–22.

Bhat, C. R. and R. Misra. 2001. Comprehensive activity travel pattern modeling system for non-workers with empirical focus on organization of activity episodes. Transportation Research Record 1777, TRB, National Research Council, Washington, DC. 16–24.

Bhat, C. R. and S. Srinivasan. 2005. A multidimensional mixed ordered-response model for analyzing weekend activity participation. *Transportation Research Part B*, 39(3): 255–278.

Bowman, J. L. and M. E. Ben-Akiva. 2000. Activity-based disaggregate travel demand model system with activity schedules. *Transportation Research Part A*, 35: 1–28.

Hamed, M. M. and F. L. Mannering. 1993. Modeling travelers' postwork activity involvement: Toward a new methodology. *Transportation Science*, 27(4): 381–394.

Horowitz, A. J. 2006. Statewide Travel Forecasting Models: Synthesis of Highway Practice. NCHRP Synthesis, 358.

Hu, P. and T. Reuscher. Summary of Travel Trends, 2001 National Household Travel Survey.

Lockwood, A., S. Srinivasan, and C.R. Bhat. An exploratory analysis of weekend activity patterns in the San Francisco Bay area, forthcoming, *Transportation Research Record.*

Parsons Brinkerhoff Quade and Douglas. Comparative analysis weekday and weekend travel with NPTS integration for the RT-HIS: Regional travel-household interview survey. Prepared for the New York Metropolitan Council and the North Jersey Transportation Planning Authority, February 2000.

Pendyala, R. M., T. Yamamoto, and R. Kitamura. 2002. On the formulation of time-space prisms to model constraints on personal activity-travel engagement. *Transportation*, 29(1): 73–94.

Strathman, J. G., K. J. Dueker, and J. S. Davis. 1994. Effects of household structure and selected travel characteristics on trip chaining. *Transportation*, 21(1): 23–45.

Handy, S. 1996. Understanding the link between urban form and nonwork travel behavior. *Journal of Planning Education and Research*, 15(3): 183–198.

Handy, S. and T. Yantis. 1997. The impacts of telecommunications technologies on nonwork travel behavior. Report No. SWUTC/97/721927-1, Southwest Region University Transportation Center, University of Texas–Austin.

Hu, P. and J. Young. 1999. Summary of travel trends: 1995 nationwide personal transportation survey. Prepared for the U.S. Department of Transportation Federal Highway Administration. Available online: http://www-cta.ornl.gov/npts (accessed October 30, 2000).

Vickerman, R. A. 1994. Demand model for leisure travel. *Environment and Planning A*, 6: 65–77.

Boarnet, M. G. and S. Sarmiento. 1998. Can land use policy really affect travel behavior? A study of the link between non-work travel and land use characteristics. *Urban Studies*, 35(7): 1155–1169.

12

TRANSFERABILITY OF TIME-OF-DAY CHOICE MODELING FOR LONG-DISTANCE TRIPS

Xia Jin
Cambridge Systematics, Inc.

Alan Horowitz
Department of Civil Engineering and Mechanics
University of Wisconsin–Milwaukee

12.1 INTRODUCTION

Time-of-day modeling deals with the times at which travel occurs throughout the day. This chapter presents a study in time-of-day choice modeling for long-distance trips with special interest in the transferability of the model. Two data sets from the 2001 National Household Travel Survey (NHTS) and the 2001 California Statewide Household Travel Survey were employed to explore the effects of various factors on time-of-day choice-making and to test the transferability of the behavioral findings and model parameters. Although there are remarkable differences in data composition between the data sets, comparative analysis of the models developed from the two data sets reveals consistent results, suggesting the potential for transferability of the behavior pattern across spatial locations.

During the past two decades, it has become increasingly apparent that it is necessary to incorporate the temporal nature of trip-making into the demand modeling process. The need to forecast traffic throughout the day has been motivated by the emerging issues that require detailed temporal resolution of trip-making, such as estimating vehicular emissions and air quality, because vehicle speeds and warm-up characteristics vary widely by time of day; assessing the effectiveness of time-of-day-specific congestion management programs, that is, congestion pricing; and evaluating the effect of travel demand management strategies on spreading peak travel times.

A fair amount of research has been conducted and various approaches proposed for time-of-day modeling. At an aggregate level, there are simple time-of-day factors and peak-spreading procedures (Arnott et al., 1994) and equilibrium-based models that account for interactions between network supply and demand (Bhat and Steed, 2002; Bhat et al., 2004; BTS, 2004; Cambridge Systematics, 1997). At a disaggregate level, choice behavior modeling has been used to examine the underlying causality of individual time-of-day choice (Ettema and Timmermans, 2003; Ettema et al., 2004; Gadda et al., 2007; Hague Consulting Group, 2000; Hess et al., 2005; Jin and Horowitz, 2008). Time-of-day modeling is also an important element in the activity-based framework that involves comprehensive daily travel activity scheduling (Kitamura et al., 2000; Okola, 2003; Stada et al., 2002; Steed and Bhat, 2000; Tringides et al., 2004; Van der Zijpp and Lindveld, 2001).

While time-of-day modeling has been drawing considerable attention in research, nearly all of the studies have focused on urban daily trips. It can be argued that choice behavior for long-distance trips is far more complicated than that for urban trips, because most urban trips are made for daily routines whereas long-distance trips are occasional and exceptional. Time-of-day analysis for long-distance trips will not only improve the forecasts of rural and intercity travel, but also supplement urban forecasts for vehicular emission and traffic congestion.

This chapter presents a study in time-of-day choice modeling for those long-distance trips with special interest in the transferability of the model. Transferability is critical to assessing the validity of behavioral models. As travel demand is derived from the need to pursue activities at different locations, travel behavior can be altered by changing spatial and socioeconomic structures. Transferability can be defined as the extent to which a model developed in a particular specific spatial, transportation, and institutional context can be applied to another context. In a previous study, a multinomial logit (MNL) model was developed to examine the time-of-day choice behavior on the 2001 NHTS data (Vovsha and Mark, 2004). The general methodology for developing the model was applied to the 2001 California Statewide Household Travel Survey data to test the transferability of the behavior findings and model parameters.

12.2 LONG-DISTANCE TRIPS

This section provides general background information about the characteristics of long-distance trips as determined from the latest nationwide survey on long-distance travel: 2001 NHTS (Zeid, 2006). Of all long-distance trips made in the United States in 2001, fewer than half (43 percent) were made by women, the majority (approximately 55 percent) were made by individuals living in households with annual household incomes of $50,000 or more, and nearly two-thirds of all long-distance trips were made by persons aged 25 to 64. About 90 percent of the long-distance trips were made by personal vehicle, with the rest taken by air (7 percent) and bus or train (3 percent). Nearly all trips (97 percent) shorter than 300 round-trip miles were made by personal vehicles. For shorter daily trips, in comparison, there were no differences across genders in the total number of trips, and daily trips showed a smaller mode share for personal vehicles (about 87 percent).

An interesting finding from the NHTS data is that nearly one of 200 commuting trips was more than 50 miles each way. Of commute trips between 50 and 99 miles, two-thirds occurred at least four days a week. These findings point to the need for more attention to long-distance trips. While not as frequent as urban trips, long-distance trips are a critical factor affecting congestion and vehicular emission issues, because 90 percent of long-distance trips are made by personal vehicles and long-distance trips account for approximately one-third of the total person-miles traveled in the United States. A final factor, which would directly affect congestion management programs, is that considerable portions of long-distance trips occur in urban areas.

12.3 DATA DESCRIPTION

The first set of data used for this analysis was extracted from the 2001 NHTS daily-travel survey records of trips 50 miles or longer in distance and taking 60 minutes or longer. For the second set of data, from the 2001 California Statewide Household Travel Survey, long-distance trips were defined as 60 minutes or longer. Information on trip distance was not available for the California data set. Trips made by travelers younger than 16 were removed from both data sets. Sixteen years was used as the break point because that is the legal age for a driver's license in California.

In total, 3,322 and 4,527 long-distance trips were identified from the NHTS and California data sets, respectively. Table 12.1 provides some basic characteristics of travel activities in the data. Compared to the NHTS sample, households in the California data had slight differences in size, number of workers, and number of vehicles, but on average made far fewer daily trips per day. Households in

Table 12.1 Comparison of basic characteristics of NHTS and California data

Measurements	NHTS	CA
Total households	1,924	2,795
Average HH size	2.84	2.74
Average number of workers per HH	1.58	1.45
Average number of vehicles per HH	2.5	2.44
Total long distance trips	3,322	4,527
Average daily trip rate per HH per day	12.19	8.69

California made slightly fewer long-distance trips on average, a difference that reasonably cannot be attributed to the looser definition of a long-distance trip used in the California data.

The two data sets were prepared in a consistent way. The same categorization method was applied to both, since they have detailed and only slightly different classifications for mode, purpose, household income, and so on. The California data, however, included no variable that indicated household structure. The entire day is aggregated into six time periods: early morning (0:00-6:29), morning peak (6:30-8:59), morning off-peak (9:00-11:59), afternoon off-peak (12:00-15:59), afternoon peak (16:00-18:29), and evening (18:30-23:59).

The two data sets differed markedly in departure time for long-distance trips (Figure 12.1). The NHTS data showed fewer trips during the two traditional peak periods than the morning off-peak period, whereas the California data showed the opposite. There may be several reasons for the difference. The California sample might have greater shares of work and/or school trips or return home trips, which are more likely to happen during the traditional peak periods. In addition, if the two data sets had different shares for weekday and weekend trips, the distributions by departure time might vary.

Both samples have about the same portion of work and/or school trips (about 26 percent), but the California data captured a larger share of return home trips than the NHTS data (37.5 percent compared to 10.9 percent). Figure 12.2 shows the breakdown of trip purpose by departure time. The bars represent the percentages of work/school and return home trips by each time period in the NHTS data, whereas the lines represent similar shares for the California sample. For work/school trips the two data sets had a consistent pattern, although work/school trips in the morning peak periods had a slightly larger share in the California sample than in the NHTS sample. A noticeable difference is that more than 50 percent of

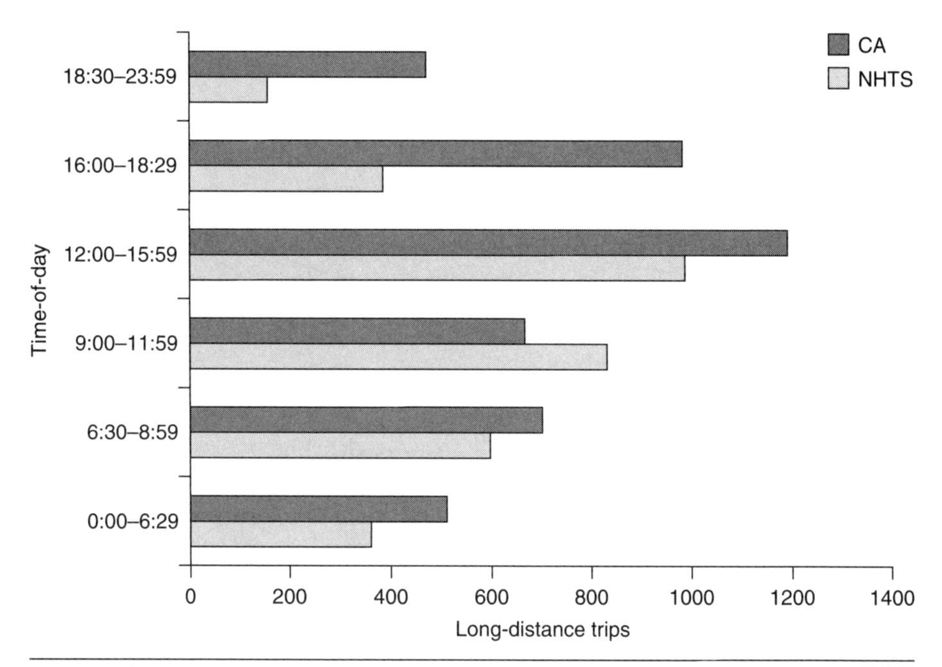

Figure 12.1 Distribution of departure time

return home trips from the NHTS sample took place during the afternoon non-peak hours, and only about 20 percent were taken in the afternoon peak period; in the California sample, about 38 percent and 33 percent, respectively, of return home trips occurred in the afternoon peak and non-peak hours. This difference may be partly due to the California sample having a much smaller share of weekend trips (only about 3.5 percent) than the NHTS sample (about 31 percent). Apparently return home trips during afternoon peak hours are more likely to take place on weekdays than during weekend trips.

The preceding analysis indicates the influence of trip purpose and travel day type on the departure time choice. On the other hand, it could also be argued that the differences in departure-time distribution in the two data sets are partly due to the varying sampling composition in trip characteristics. Whether there are distinctions in choice behavior under different geographical and socioeconomic contexts cannot be established at this point based on simple classifications. MLN modeling analysis, a more sophisticated method of looking at behavioral patterns, would provide some insights in this perspective, which we discuss later in this chapter.

Figures 12.3 and 12.4 are prepared in a similar fashion to Figure 12.2. Figure 12.3 presents a comparison between the two data sets in departure time distribution by travel day type. As for weekend long trips, in the California sample fewer

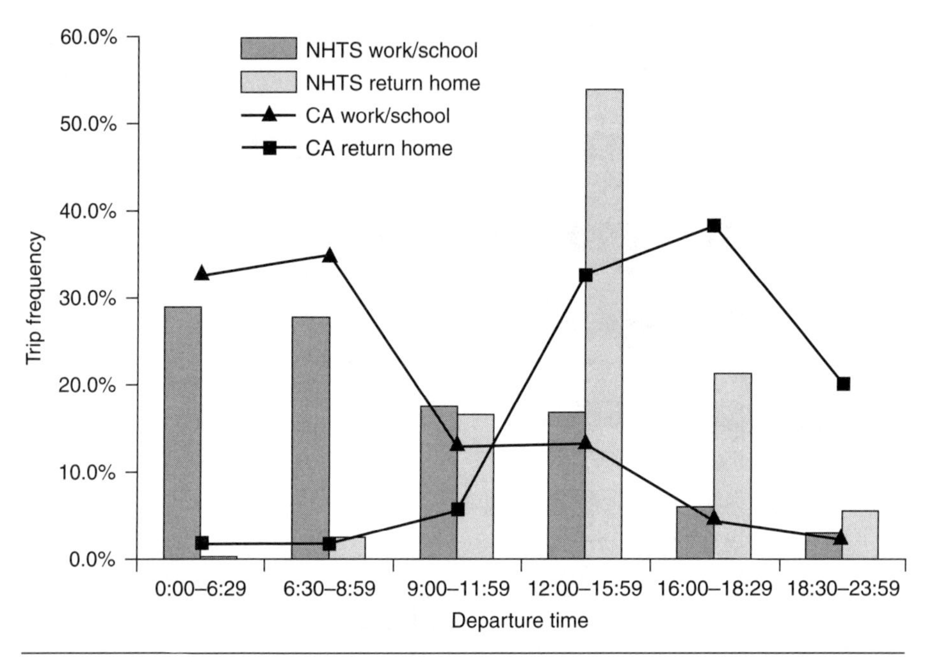

Figure 12.2 Distribution of departure time by purpose

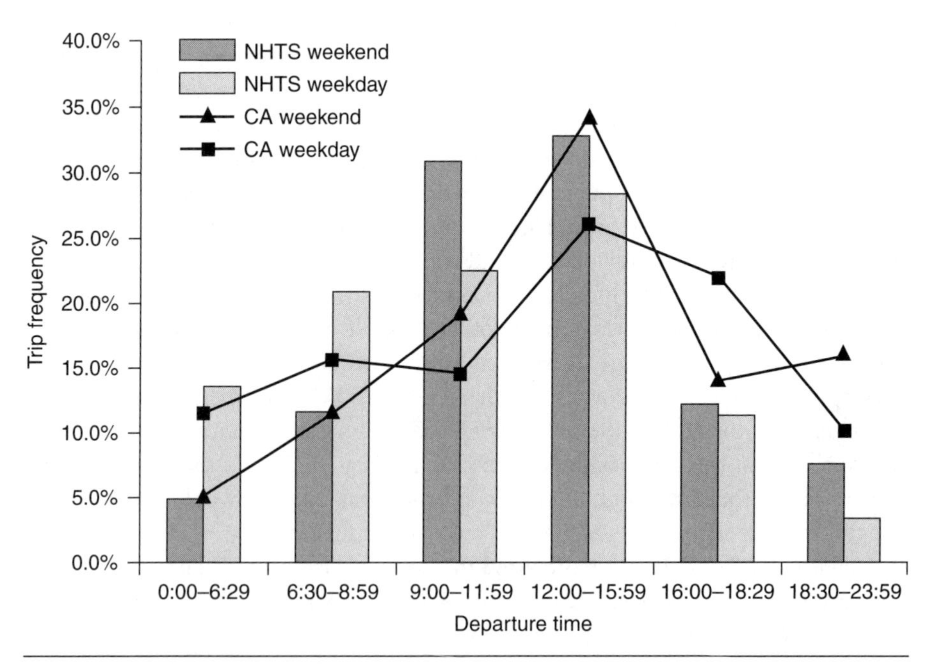

Figure 12.3 Distribution of departure time by travel-day type

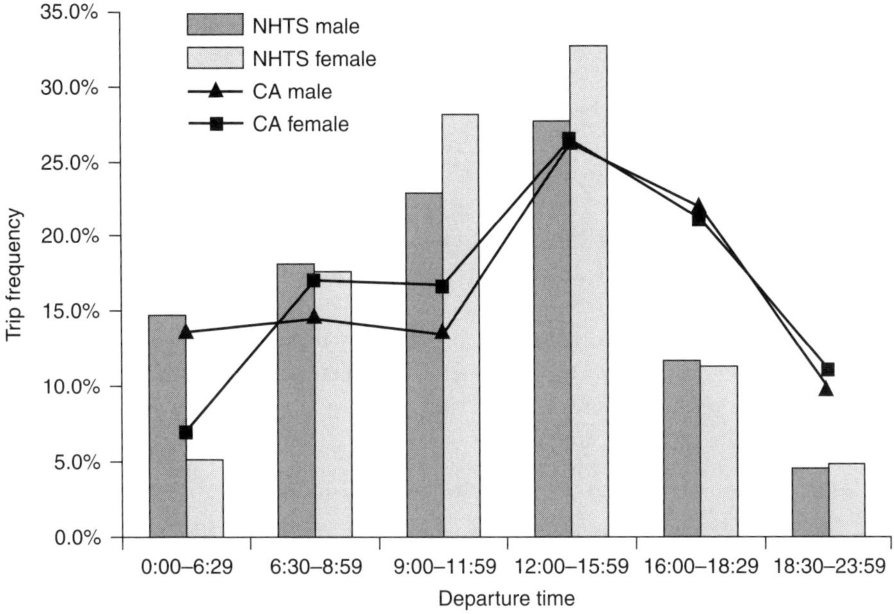

Figure 12.4 Distribution of departure time by gender

trips were taken in the morning off-peak period and more trips in the evening hours than in the NHTS sample. Interestingly, in the NHTS sample, the share in the afternoon peak period was higher during weekend trips than weekday trips.

Figure 12.4 shows the distribution of departure time by gender for both data sets. In general, the two data sets showed consistent patterns in comparing the departure time choices of men and women. Males made more long trips in early morning and fewer trips in morning off-peak hours than females. Again, households in the California sample made more long-distance trips in the afternoon peak period and fewer trips in the morning off-peak periods, compared to the NHTS sample. The difference can be partly explained by the California sample having a larger share of weekday trips, but MNL analysis can provide more robust reasoning.

12.4 MNL MODELING ANALYSIS AND RESULTS

12.4.1 MNL Modeling

MNL modeling provides the opportunity to examine the effects of various factors on long-distance trip time-of-day choice and to better understand the underlying causality of the choice behavior. The procedure forecasts the probability of one time of day (TOD) being chosen for each long-distance trip based on a set of taste parameters and the attributes of the alternatives and the decision maker. The model used in this study is described:

$$\Pr(TOD) = f(T, A, P, H) \tag{12.1}$$

where:

TOD = TOD period choice for the long trip
T = trip-related factors such as purpose, mode, travel time, traveling companions
A = activity related factors such as activity duration
P = personal characteristics such as age, gender, education level
H = household characteristics such as income, size, auto-ownership, presence of young child.

The last time period (18:30-23:59) served as the reference category for interpreting the MNL model, because long-distance trips were least likely to happen in this time period. Among the many results, an MNL model provides a separate estimate of the probability of departing at each of the time periods relative to the evening period. Similarly, for each categorical independent variable the last category served as the base category, so the coefficients were calculated relative to these arbitrary base categories.

12.4.2 Model Specifications

The initial model developed on the 2001 NHTS data was modified to exclude trips made by non-automobile modes, because all long-distance trips extracted from the 2001 California data were made by automobiles. Various types and combinations of explanatory variables were examined in MNL analysis to determine whether and to what extent these variables affect time-of-day choice for long-distance trips.

Table 12.2 presents the factors tested in the models and the final sets of model specifications being chosen. As shown in the table, some variables played an important role in the choice behavior and were irreplaceable; some variables had influence on the decision making and could be included in the model when other related variables were not available; and the other variables did not reveal any significant relationship with the departure-time choice in all three models. In general, the behavioral findings are consistent across both NHTS models and the California model. The main difference is that the number of non-household members in travel was significant in the NHTS models, whereas the number of household members in travel was significant in the California model, and the two variables were not interchangeable in any model.

Table 12.2 Various factors tested in MNL models

	NHTS all modes	NHTS auto only	CA auto only
Activity duration	▲	▲	▲
Trip duration	▲	▲	▲
Purpose	▲	▲	▲
Weekend	▲	▲	▲
Mode	△	–	–
Driver or passenger	×	▲	△
HH vehicle used	×	▲	×
Total number of people in travel	×	×	△
Number of HH member in travel	×	×	▲
Number of non-HH member in travel	▲	▲	×
Travel was overnight	×	×	×
Gender	▲	▲	▲
Age	▲	▲	▲
Education level	▲	▲	△
Employed	▲	▲	▲
Multi-job status	△	△	△
HH income	×	×	△
HH size	△	△	×
Number of HH worker	▲	▲	△
Number of HH vehicle	×	×	▲
Number of HH student	–	–	×
HH life cycle	▲	▲	–
Housing unit owned or rented	×	×	×
Household in urban or rural area	×	×	×
# of cases	3322	3106	4524
−2 log likelihood – intercept only	10943.245	10190.167	15699.053
−2 log likelihood – final	9325.567	8668.975	12807.416
R square	0.400	0.402	0.487

▲ significant and chosen in model   × not significant
△ significant at a lesser degree   – N/A

12.4.3 Model Parameter Coefficients

The logit coefficient is used to describe the effect of a single variable on predicting the dependent variable—in this study, travelers' time-of-day choice. To explain the results of logit regression more effectively, another form of logit regression—exp(B) —was used:

$$\text{odds}(p) = \frac{p}{1-p} = \exp(a + b_1x_1 + b_2x_2 + b_3x_3 + \cdots) \tag{12.2}$$

The left side of the equation indicates the odds ratio (the probability of something being true divided by the probability of it not being true) for choosing a specific departure time period; the right side is the exponential of the regression predictor. Holding other variables constant, when the independent variable x_i increases by one unit, the odds that the dependent variable is equal to a chosen time period increases by a factor of exp(B).

The NHTS all-modes model has been previously presented (Jin and Horowitz, 2008) and is not included in this chapter. The estimated results for the NHTS auto-only model and the California model are presented in Table 12.3 and Table 12.4. The numbers in gray indicate that the coefficients were not statistically significant at 0.2 level.

The two MNL models are consistent in their behavioral interpretations, with a few distinctions at a fine level. The findings from the MNL models are summarized as follows:

- Trip duration, activity duration, trip purpose, and whether the trip took place on a weekend were the most powerful factors in determining the TOD choice for long-distance trips. The longer the trip itself or the longer the time to be spent at the destination, the earlier the trip would be taken. Compared to weekday long trips, weekend long-distance trips had a higher possibility of taking place in the evening period, followed by the midday periods.
- Travelers taking different types of long trips exhibited various preferences on the departure time. Work or related trips most often took place in the early morning and morning peak periods. Return home trips had the highest probability of occurring in the afternoon periods as indicated by the NHTS models, and in the evening period as shown in the California model. Travelers from the NHTS sample were more likely to depart in the morning for personal business trips and in the afternoon for recreation trips. Travelers from the California study were more likely to depart during the midday and the evening.
- Travel mode, in general, was not statistically significant to the TOD choice in the original NHTS analysis. However, for long automobile

Table 12.3 Parameter coefficients for NHTS—auto model

		0:00–6:29	6:30–8:59	9:00–11:59	12:00–15:59	16:00–18:29
N_NonHH		0.848	0.914	0.803	0.864	1.002
Trip duration		1.014	1.010	1.007	1.004	1.000
Activity duration		1.017	1.014	1.011	1.009	1.007
Weekend (1 if weekend)		0.295	0.283	0.527	0.421	0.357
Sex (1 if male)		1.540	0.847	0.744	0.931	1.296
Dr_Pass (1 if driver)		0.830	0.883	1.070	0.925	0.626
hhveh_used (1 if used)		2.021	1.988	1.279	1.213	1.442
Worker (1 if worker)		0.827	0.592	0.588	0.514	0.883
Purpose	Work/school	7.177	4.608	1.992	1.475	0.727
	Return	0.042	0.190	0.712	1.972	1.341
	Persn busin	1.280	1.954	1.558	1.157	0.693
	Soci recrt	0.521	0.930	1.300	1.274	0.691
	Others	—	—	—	—	—
Education	HS or less	4.694	1.968	1.520	1.009	1.086
	HS	2.284	1.271	1.168	1.041	1.038
	College	3.498	2.358	2.129	2.162	1.466
	Graduate	—	—	—	—	—
LifeCycle	No child	0.543	0.882	0.884	1.057	1.119
	Child 0-5	0.489	0.499	0.724	0.709	0.542
	Child 6-15	0.453	0.519	0.815	0.995	0.723
	Child 16-21	0.304	0.568	0.683	0.757	0.705
	Retired	—	—	—	—	—

trips, whether the traveler drove or rode along, had an impact on the departure time choice. Drivers had greater chances of making long trips in the morning off-peak period and fewer chances in the afternoon peak hours compared to trips with passengers.

- Generally, traveling with other people, whether with household members or non-household members, would increase the probability of making long-distance trips in the evening.
- The travelers' age, gender, work status, and education level all presented significant impacts on the TOD choice for long-distance trips. Older travelers would depart for long trips more often in the morning. Consistent with the NHTS models, the California model indicates

Table 12.4 Parameter coefficients for California model

		0:00–6:29	6:30–8:59	9:00–11:59	12:00–15:59	16:00–18:29
Num_HH		0.606	0.755	0.906	0.857	0.789
Trip duration		1.008	1.006	1.007	1.004	1.000
Activity duration		1.004	1.002	1.001	1.001	1.001
Weekend (1 if weekend)		0.373	0.565	0.767	0.887	0.414
Sex (1 if male)		1.905	0.914	1.026	1.187	1.140
Age		1.020	1.021	1.034	1.020	1.010
HHveh		1.007	0.904	0.873	0.914	0.954
Purpose	Work/school	4.538	5.675	3.073	1.550	0.533
	Return	0.007	0.013	0.072	0.267	0.383
	Persn busin	0.916	1.798	2.306	2.103	1.458
	Soci recrt	0.187	0.534	0.947	0.683	1.048
	Other	—	—	—	—	—
Work status	Employed	3.229	2.224	0.545	1.755	3.738
	Work home	1.898	2.935	1.112	2.106	3.303
	Non-worker	3.349	3.361	1.235	2.628	2.255
	other	—	—	—	—	—

that males were more likely to make long-distance trips in the early morning period than females. The California model shows that males had a higher possibility of making long trips in the midday periods relative to the evening period, whereas the NHTS models indicated the opposite. Workers had less probability of departing for long trips in the midday periods than non-workers.

- Household characteristics, including household income, household size, number of workers, and number of vehicles revealed various impacts on the TOD choice for long-distance trips, depending on the model. Only the number of household workers showed significant effects in all models.

12.5 SUMMARY

The 2001–2002 NHTS and the 2000–2001 California Statewide Household Travel Survey provided good data to examine the time-of-day choice behavior for long-distance trips and to test the possible spatial transferability of the behavior findings and models. A wide range of factors was explored, and the results indicated that departure-time choice-making behavior for long, occasional, and exceptional trips is more complicated than for urban short trips. Trip and activity duration, trip purpose, travel day type, and various personal and household characteristics all exhibited significant relationships with departure-time choice-making. A closer look at the travel demands at certain times of the day would help evaluate the impacts on the performance of the transportation network more effectively, and develop appropriate improvement programs to serve regional or statewide transportation needs. Although there are remarkable differences in data composition between the two data sets, comparative analysis of the models developed from the two data sets reveals consistent results, suggesting the potential for transferability of the behavioral pattern across spatial locations.

REFERENCES

Arnott, R., A. de Palma, and R. Lindsey. 1994. Welfare effects of congestion tolls with heterogeneous commuters. *Journal of Transport Economics and Policy*, 28(2): 139–161.

Bhat, C. R. and J. L. Steed. 2002. A continuous-time model of departure time choice for urban shopping trips. *Transportation Research*, 36B, 3: 207–224.

Bhat, C. R., J. Y. Guo, S. Srinivasan, and A. Sivakumar. 2004. A comprehensive econometric micro-simulator for daily activity-travel patterns. CEMDAP.

BTS. 2004. National household travel survey long-distance travel quick facts. http://www.bts. gov/programs/national_household_travel_survey/long_distance.html.

Cambridge Systematics. 1997. Time-of-day modeling procedures state-of-the-art, state-of-the-practice. http://tmip.fhwa.dot.gov/clearinghouse/docs/Time-Day/time-day.pdf

Ettema, D. and H. Timmermans. 2003. Modeling departure time choice in the context of activity scheduling behavior. *Transportation Research Record 1831*. Washington, DC. TRB, National Research Council, 39–46.

Ettema, D., O. Ashiru, and J. W. Polak. 2004. Modeling timing and duration of activities and trips in response to road-pricing policies. *Transportation Research Record 1894*. Washington, DC. TRB, National Research Council, 1–10.

Gadda, S., K. M. Kockelman, and P. Damien. 2007. Continuous departure time models: A Bayesian approach. TRB 2007 Annual Meeting. CD-ROM.

Hague Consulting Group. 2000. Modeling peak spreading and trip re-timing. UK Department of Transport. http://www.dft.gov.uk/stellent/groups/dft_econappr/documents/pdf/dft_econappr_pdf_504841.pdf

Hess, S., A. Daly, and J. Bates. 2005. Departure time and mode choice analysis of three stated preference data sets. http://www.transport.gov.uk/stellent/groups/dft_econappr/documents/ page/ dft_econappr_040159.pdf.

Jin, X. and A. J. Horowtiz. 2008. Time-of-Day Choice Modeling for Long-Distance Trips. *Transportation Research Record*, 2076, 200–208.

Kitamura, R., C. Chen, R. M. Pendyala, and R. Narayanan. 2000. Micro-simulation of daily activity-travel patterns for travel demand forecasting. *Transportation*, 27(1): 25–51.

Okola, A. 2003. Departure time choice for recreational activities by elderly non-workers. Transportation Research Record 1848. Washington, DC: TRB, National Research Council, 86–93.

Stada, J., S. Logghe, G. D. Ceuster, and B. Immers. 2002. Time-of-day modeling using a quasi-dynamic equilibrium assignment approach. http://www.tmleuven.be/Vervoer/Paper_200203.pdf.

Steed, J. L. and C. R. Bhat. 2000. On modeling departure-time choice for home-based social/recreational and shopping trips. *Transportation Research Record 1706*. Washington, DC. TRB, National Research Council, 152–159.

Tringides, C. A., X. Ye, and R. M. Pendyala. 2004. Departure-time choice and mode choice for non-work trips: alternative formulations of joint model systems. *Transportation Research Record 1898*. Washington, DC. TRB, National Research Council, 1–9.

Van der Zijpp, N. J. and C. D. R. Lindveld. 2001. Estimation of origin-destination demand for dynamic assignment with simultaneous route and departure time choice. *Transportation Research Record 1771*. Washington, DC. TRB, National Research Council, 75–82.

Vovsha, P. S. and B. Mark. 2004. Hybrid discrete choice departure time and duration model for scheduling travel tours. *Transportation Research Record 1894*. Washington, DC. TRB, National Research Council, 46–56.

Zeid, A., T. F. Rossi, and B. Gardner. 2006. Modeling time of day choice in the context of tour and activity based models. *Transportation Research Record 1981*. Washington, DC. TRB, National Research Council, 42–49.

13

UNIVARIATE SENSITIVITY AND UNCERTAINTY ANALYSIS OF STATEWIDE TRAVEL DEMAND AND LAND USE MODELS FOR INDIANA

Li Jin
Kittelson and Associates

Jon D. Fricker
School of Civil Engineering
Purdue University

13.1 INTRODUCTION

Uncertainty exists in statewide travel demand forecasting and land use models. Both input data and adopted model parameters can vary from their true values because of model misspecification, imperfect input information, and innate randomness of events. The objective of this chapter is to study the sensitivity of vehicle miles of travel (VMT) outputs of *in*tegrated *tr*avel and *l*and *u*se *de*mand (INTRLUDE), an integrated statewide travel demand and land use forecasting model system for Indiana, to plausible amounts of variations on travel model parameters and input data. The land use in Central Indiana 2 (LUCI2) model is used in conjunction with the statewide travel demand model to predict travel demand through various scenarios over time. Results indicate that VMT outputs are most sensitive to trip distribution function parameters and trip production rates. Population growth rates and trip assignment parameters don't have a high degree of influence on VMT outputs.

Uncertainty in empirical quantities can arise for a variety of reasons, such as (Morgan and Henrion, 1990):

1. Random error and statistical variation
2. Systematic error and subjective judgment
3. Linguistic imprecision
4. Variability
5. Randomness and unpredictability
6. Disagreement
7. Approximations

Travel demand predictions involve substantial uncertainty, which can derive from model misspecification, imperfect input information, and innate randomness in events (Krishnamurthy and Kockelman, 2003). Therefore, sensitivity and uncertainty analysis are of great importance for policy and risk analysis. With explicit recognition of uncertainty in travel demand forecasts, policymakers can understand the influence of the values of model parameters and input data and, as a result, able to make more informed choices when allocating financial resources to collect input data and calibrate model parameters. Analysts can also devote greater effort to acquiring input data and adopting model parameters when the impact of a change in each of the output is known.

The Intermodal Surface Transportation Efficiency Act (ISTEA) of 1991 required that metropolitan and statewide transportation plans be integrated with land use plans (Miller et al., 1999). It would be helpful to do a statewide sensitivity and uncertainty analysis when using these complex integrated transportation and land use models. This study introduces a univariate uncertainty analysis using Indiana's statewide INTRLUDE model system. As already stated, the statewide urban simulation model, LUCI2, is used in conjunction with the statewide travel demand model to predict the travel demand over time using various scenarios.

This chapter includes the following sections: background and literature review, model description, and sensitivity simulation results. The study concludes with a summary of the research findings and a discussion of possible future research.

13.2 BACKGROUND

13.2.1 Literature Review

A few studies have been conducted to quantify uncertainty in predictions of travel demand using a four-step travel demand model or an integrated transportation-land use model system.

Zhao and Kockelman (2002) investigated uncertainty propagation in four-step travel demand models over a 25-zone network. Monte Carlo simulation and sensitivity analysis were used to quantify variability in model outputs. Their results suggested that uncertainty was compounded over the four stages of a transport model and was highly-correlated across outputs. Traffic assignment, the final step of the model, was found to reduce uncertainty developed in the first three steps, but, in general, it could not reduce final flow uncertainty below the levels of input uncertainty.

Pradhan and Kockelman (2002) studied the uncertainty propagation in an integrated land use-transportation model framework using 271 traffic analysis zones (TAZs). Urban development was modeled using UrbanSim (Waddell, 2002). A factorized design approach was employed in the present study. The authors found that, whereas several model inputs might affect model outputs in the short run, only those inputs that had a cumulative effect had the potential of having a significant impact on outputs in the long run. Their results also suggested that uncertainty in model outputs might increase for the first few years, although the level of uncertainty appeared to diminish in later years as households, jobs, and developers responded to changed input conditions.

Krishnamurthy and Kockelman (2003) examined the propagation of uncertainty for 1,074 traffic serial zones (TSZs) using Putman's integrated transportation and land-use package (ITLUP) and urban transportation planning package's (UTPP) traditional four-step travel demand model. The ITLUP consists of a disaggregate residential allocation model (DRAM) and an employment allocation model (EMPAL) (Putnam, 1983). Results indicated that output variations were most sensitive to the exponent of the link performance function, the distribution of trips between peak and off-peak periods, and several trip production and attraction rates.

A study conducted by Clay and Johnston (2005) introduced the univariate uncertainty analysis of an Integrated Land Use and Transportation Model. MEPLAN was applied to the 81 zones (including 10 external zones) of the Sacramento Area Council of Governments (SACOG) model. The study found that uncertainty in the socio-economic forecasts did not dominate the final number of errors observed in the model outputs.

Because of the high computational cost and complexity associated with statewide travel demand systems, limited research has looked at sensitivity and uncertainty analysis for statewide models. This chapter, however, investigates the uncertainty of the statewide travel demand predictions arising from the model parameters and input data involved in using an integrated statewide travel demand and land use forecasting model system for Indiana. Outputs from the LUCI2 urban simulation model act as the inputs to Indiana's statewide four-step travel demand model (ISTDM). Congested skim trees from the traffic assignment stage of the ISTDM are fed forward to the LUCI2 urban simulation model for the next five-year period.

The Land Use in Central Indiana (LUCI) urban simulation model, developed by Ottensmann (2003), was initially implemented for a 44-county region in Central Indiana. The model generated alternative development scenarios that incorporated various policy choices about future land use developments using urban development data from 1985, 1993, and 2000 that was derived from Landsat Thematic Mapper satellite images with 30-meter resolution. The LUCI model relied on random utility and discrete choice theory as the basis for predicting urbanization. The LUCI2 urban simulation model, which extends the LUCI model, covers the entire state and is integrated with the ISTDM. The integrated ISTDM and LUCI2 urban simulation model system is used in this study.

13.2.2 Calculating Univariate Uncertainty

Earlier studies used different ways to calculate uncertainty. Monte Carlo techniques had been used to draw input values from multivariate distributions (Krishnamurthy and Kockelman, 2003; Zhao and Kockelman, 2002). Because the multivariate distributions were not known to researchers, they were normally chosen by unverified assumptions. Clay and Johnston (2005) used a different technique to compute the univariate uncertainty: they varied exogenous production rates, commercial trip generation rates, and a concentration parameter by ± 10 percent, ± 25 percent, and ± 50 percent per source, based on careful review of previous studies and consultation with the firm that calibrated the model.

In the present study, we calculate uncertainty using the method employed by Clay and Johnston. Given unknown probability distributions of model parameters and input data, the moment estimations of output distribution will be biased if the assumed input distribution does not closely approximate the actual one. Also, because the Monte Carlo technique requires large sample sets to cover the possible ranges of all inputs, and consequently a long computing time for each run, it is difficult to apply the Monte Carlo technique to integrated statewide models. In this study, therefore, only one set of the model parameter or input data varies by ± 10 percent per run. The variations of ± 10 percent to the true value are considered to occur in practice with high probability for all the model parameters and input data. The variations of 25 percent and 50 percent are not applied because only small and realistic variations of the model parameters and input data are of interest for this chapter.

In each simulation run, keeping all the other model parameters or input data the same as the base case, the value of one set of model parameters or input data is changed by ±10 percent. The statewide VMT outputs are then compared with the base case (with all the model parameters or input data unchanged) and the results are analyzed.

ISTDM and LUCI2 exchange input/output data for each five-year period. The simulation is set to run for the period 2000 to 2030. A total of 53 model runs were performed in this study with each taking about eight hours to finish on the computer with Intel Core2 Duo 1.80GHz and 1GB 667MHz RAM.

13.3 MODEL DESCRIPTION

13.3.1 Data Description

The Indiana statewide travel demand model has 4,720 TAZs (including 141 external TAZs) and 35,300 roadway links. It encompasses all 92 counties in Indiana and parts of neighboring states (Bernardin, Lochmueller and Associates; Cambridge Systematics, 2005). Figure 13.1 shows the ISTDM model area.

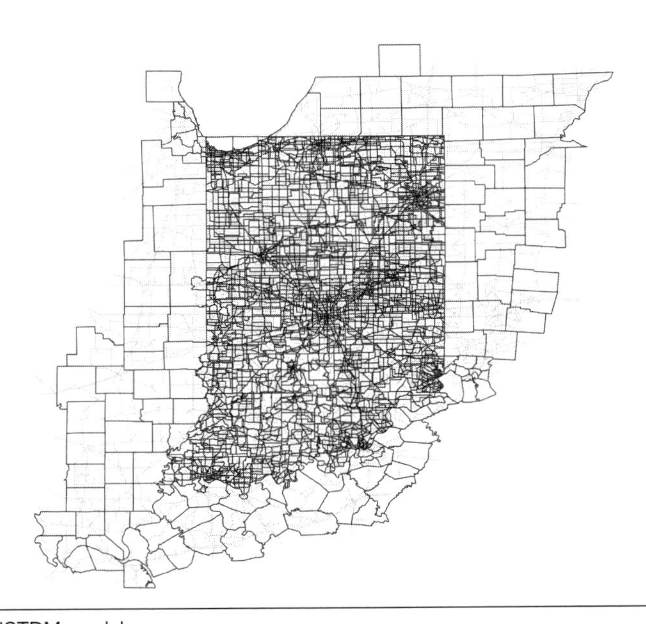

Figure 13.1 ISTDM model area

INDOT's new road inventory data (RID) for 2000 were attached to the network. The 1995 Indiana Travel Survey and 2000 census data were used for the Indiana statewide four-step travel demand model. Satellite land cover data for 1985, 1993, and 2000 were used for the LUCI2 urban simulation model.

13.3.2 LUCI2 Urban Simulation Model Estimation

The LUCI2 urban simulation model predicts changes in employment and conversion of nonurban land to residential and employment-related uses. The simulation model uses random utility theory, with aggregate logit discrete choice models for converting available nonurban land to residential and employment-related land uses (Ottensmann, 2007).

13.3.2.1 Residential Allocation Model

The residential development model of the LUCI2 urban simulation model includes two parts: (1) probability of residential development model and (2) density of residential development model. The formula for the probability of residential development model is:

$$\text{logit}(p_i) = \log\left(\frac{P_i}{1-p_i}\right) \tag{13.1}$$

where p_i is the proportion of land converted to residential use. The equation for the aggregated logit model is:

$$\text{logit}(p_i) = \beta_0 + \sum_k \beta_k X_{ik} \tag{13.2}$$

where logit(p_i) is the logit proportion of converted land from 1995 to 2000 (reduced by 0.625 from 1993 to 2000); X_{ik} is the matrix of independent variables that include accessibility to employment in 1995; sewer service dummy, proportion of residential land in the simulation zone and its square in 1993; and logit proportion of land converted to residential from 1985 to 1993.

The accessibility to employment in 1995 is calculated using the following formula:

$$A_i = \sum_j E_j e^{\beta T_{ij}} \tag{13.3}$$

where E_j is employment in TAZ j, T_{ij} is the congested travel time from TAZ i to TAZ j (which is calculated by the Indiana statewide four-step travel demand model), and β is an empirically determined accessibility coefficient.

The formula for the population density of new residential development is a multiple regression model:

$$\log(D_i) = \beta_0 + \sum_k \beta_k X_{ik} \tag{13.4}$$

where $\log(D_i)$ is the natural logarithm of population density in 2000, X_{ik} is the matrix of independent variables, including the log accessibility to employment using year 2000 INDOT employment, sewer service dummy, and the natural logarithm of the percentage of students passing the Indiana Graduation Qualifying Examination in 2002. The accessibility is calculated as Equation (13.3) with different parameters β.

Urban development in the model is driven by an exogenous forecast population growth input for the entire region. Users specify the population growth from TransCAD GISDK interface when running the integrated transportation-land use model. In this study, the 1990–2000 population growth rate of 4.725 percent over five years is used as the base case. The uncertainty of population growth rate is assigned with ± 10 percent for the simulation runs.

13.3.2.2 Employment Allocation Model

For the employment development model, the first step is the prediction of employment changes by industry for each TAZ zone. This is a multiple regression model as shown in Equation (13.4). The dependent variable is the change in employment in service industries from 1995 to 2000. The independent variables include the change in accessibility to population from 1990 to 1995, population in TAZs in 1995, population change in TAZs from 1990 to 1995, change in urban land in TAZs from 1985 to 1993, percent of land area as employment-related land use in 1993, and percent of the land area in the TAZ available for urban development in 1993. The accessibility is calculated using Equation (13.3) with different parameters β.

The next stage is the simulation of employment-related development for each of the simulation zones. Two models are developed: the first predicts employment density (employment-related land required per employee) for each TAZ; the second predicts the probability of land conversion to employment-related land use. The formula for the prediction of employment density is a multiple regression model as seen in Equation (13.4). The dependent variable is the natural logarithm of employment-related land use per employee in 2000. The independent variables include the housing units built before 1940, population change from 1995 to 2000, and percent of urban land in 2000. The formula for the prediction of the probability of land conversion to employment-related land use is an aggregate logit model as shown in Equations (13.1) and (13.2). The dependent variable is the logit proportion of increase in the amount of employment-related land from 1995 to 2000 (reduced by 0.625 from 1993 to 2000). The independent variables include the log accessibility to population, sewer service dummy, and employment-related land use in 1993. The accessibility is calculated using Equation (13.3) with different parameters β.

13.3.3 Indiana Statewide Travel-Demand Model

INTRLUDE is a four-step travel demand model that links the LUCI2 urban simulation land use model to the statewide roadway network. The trip purposes considered in this sensitivity study are home-based work (HBW), home-based other (HBO), and non home-based (NHB) trips. Table 13.1 presents all the variations of the model parameters and input data for this sensitivity study. It does not include the base case run, which keeps all model parameters at their original values.

13.3.3.1 Trip Generation

Trip production rates are developed based on the observed data from the 1995 Indiana travel survey and the 2001 NHTS (Bernardin, Lochmueller and Associates; Cambridge Systematics, 2005). They are cross-classified by household size and auto ownership within each area type (urban, suburban, and rural) for each trip purpose (HBW, HBO and NHB). The uncertainty of trip production rates for each area type and trip purpose is assigned to be ± 10 percent during the simulation runs.

Trip attractions are defined as:

$$\begin{aligned}
A_{HBW} &= f(\text{Employment}_{retail,etc.}, \text{Employment}_{non-retail,etc.}) \\
A_{HBO} &= f(\text{Employment}_{retail}, \text{Employment}_{fire,etc.}, \text{Employment}_{education}, \text{Households}) \\
A_{NHB} &= f(\text{Employment}_{retail}, \text{Employment}_{fire,etc.}, \text{Employment}_{non-retail,etc})
\end{aligned} \tag{13.5}$$

where A_{HBW}, A_{HBO}, and A_{NHB} are the trip attractions for HBW, HBO and NHW. $\text{Employment}_{retail}$ is defined as the number of retail jobs in the corresponding zone.

Table 13.1 Input changes for simulation runs

Model	Model parameters	Uncertainty	Simulation runs
Land use	Population growth rate	±10%	2
Trip production	Trip production rates for each area type and trip purpose	±10%	18
Trip attraction	Trip attraction rates for each trip purpose	±10%	6
Trip distribution	Gamma function parameter *b* or *c*	±10%	12
Trip assignment	Link performance function *a* and *b*	±10%	14

The range of trip attraction rates for each trip purpose is assigned to be ±10 percent during the simulation runs.

13.3.3.2 Trip Distribution

The statewide model uses the gravity model for trip distribution. The production-constrained gravity model is defined as:

$$T_{if} = P_i \left(\frac{A_j F(t_{if})}{\sum_k A_k F(t_{ik})} \right) \tag{13.6}$$

where T_{ij} is the number of trips from zone i to zone j, P_i is the number of trip productions in zone i, A_j is the number of trip attractions in zone, t_{ij} is the travel time from zone i to zone j, and $F(t_{ij})$ is the impedance function between zone i and zone j.

The key component of the gravity model is the friction factor function. The factors were calibrated to the observed travel times obtained from the 1995 Indiana travel survey in ISTDM. To make the sensitivity analysis feasible, the gamma function distributions are used to fit the observed travel times as follows:

$$F(t_{if}) = t_{if}^{\ -b} e^{-c^{*} t_{if}} \tag{13.7}$$

Figures 13.2-13.4 show the nonlinear fitting for HBW, HBO, and NHW trips by the gamma functions.

For the simulation runs, the assigned value of b or c in the fitted gamma functions for each trip purpose is ±10 percent.

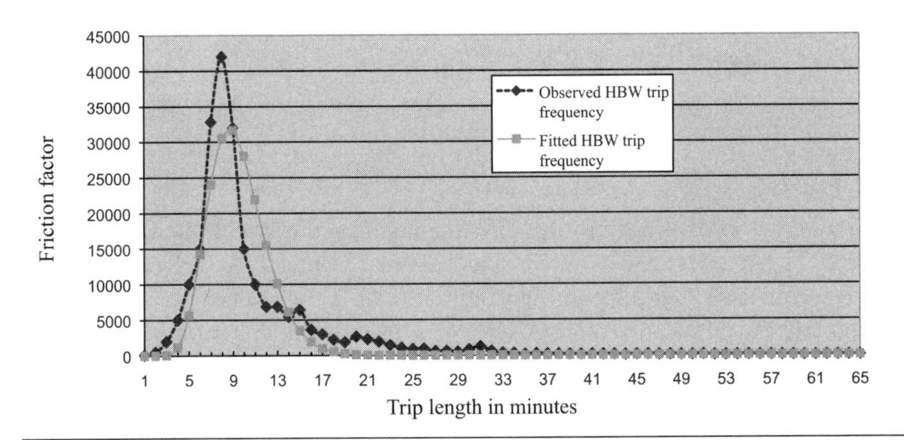

Figure 13.2 Friction factors for HBW trips

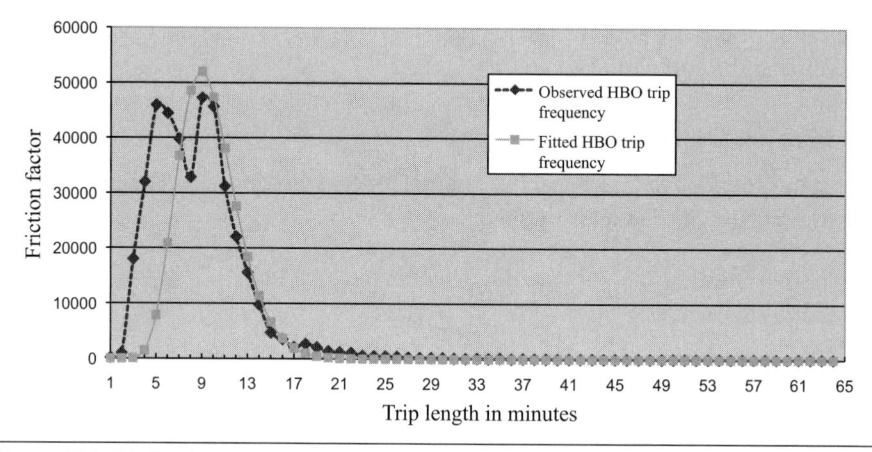

Figure 13.3 Friction factors for HBO trips

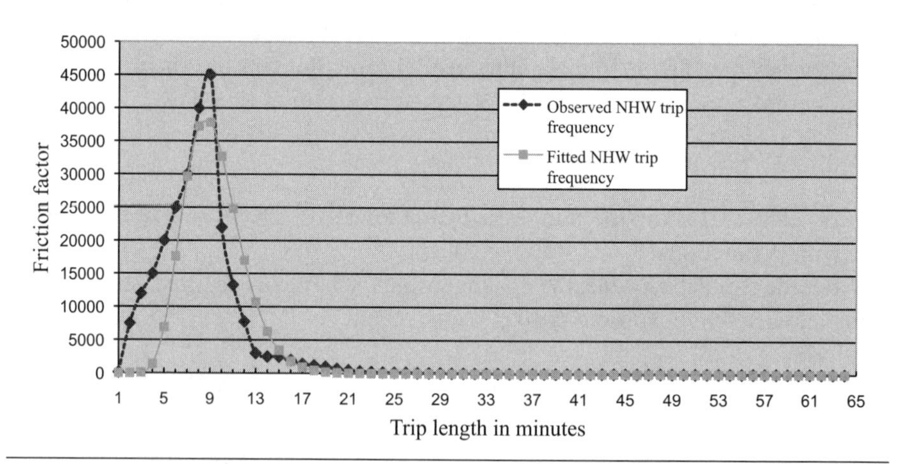

Figure 13.4 Friction factors for NHW trips

13.3.3.3 Mode Split

The mode shares developed for ISTDM are based on the 1995 Indiana household survey. They are extracted as a table and classified by area type (urban, suburban, or rural) for each trip purpose. No multinomial logit models are used for the shorter trip purposes (HBW, HBO, NHW). Multinomial logit model is only used for long trips that have a trip length more than 56 minutes. The sensitivity analysis of mode shares is not investigated in this study, because no model share parameters for shorter trip purposes (HBW, HBO, NHW) exist in the ISTDM.

13.3.3.4 Trip Assignment

The ISTDM employs a simultaneous multimodal multiclass assignment (MMA) method for trip assignment using the user equilibrium principle over a 24-hour period. The maximum iterations are 200, and the convergence criterion is set at 0.0050. Toll impedances are converted to travel times, and the trip assignment is based on travel times. Multiple volume-delay functions (VDF) are used by functional classification on the basis of extensive experimentation during model validation (Bernardin et al., 2005). The Bureau of Public Roads volume-delay function is defined as:

$$t = t_f\left[1 + a\left(\frac{V}{C}\right)^b\right] \tag{13.8}$$

where t is the travel time of a given link at traffic flow V, t_f is the free flow travel time, C is the link capacity, and a and b are volume delay function coefficients.

Table 13.2 shows different a and b by the functional class. The variations of a and b for the functional class are assigned the value of ± 10 percent during the simulation runs.

13.4 SIMULATION RESULTS

In this study, sensitivity and uncertainty analyses are processed by varying one set of the model parameters and input data per simulation run. The outputs of VMT are compared with the VMT output from the base case. Table 13.3 presents the

Table 13.2 BPR function coefficients by functional class

Functional class	a	b
1	0.71	2.1
2	0.71	2.1
6	0.71	2.1
7	0.88	9.8
8	0.88	9.8
11	0.83	5.5
12	0.71	2.1
14	0.88	9.8
16	0.56	3.6
17	0.56	3.6
19	0.56	3.6

Table 13.3 VMT outputs from the base case

Year	VMT	Increase from year 2000 (%)
2000	92950222.46	0
2005	97931196.38	5.36
2010	106059808.21	14.10
2015	114630216.69	23.32
2020	123388016.04	32.75
2025	132482531.12	42.53
2030	142285147.38	53.08

VMT from the base case, using all the original model parameters or input data and the default population growth rate of 4.725 percent over five years.

Figure 13.5 shows the percent change in the VMT outputs caused by the different levels of exogenous population growth rate compared with the base case in the corresponding year.

Because the population growth rate affects traffic patterns for future years only, the VMT outputs from the different levels of population growth rate are the same as the base case in model year 2000. In model year 2005, the VMT output increases 0.23 percent when the population growth rate increases 10 percent between 2000 and 2005, and decreases 0.29 percent when the population growth rate decreases 10 percent in the same period. In model year 2030, the VMT output increases 1.62 percent when the population growth rate increases 10 percent for each five-year period between 2000 and 2030, and decreases 1.54 percent when the population growth rate decreases 10 percent for each of the same five-year periods. The sensitivity effect on the VMT outputs increases or decreases with the model year because the growth rate increases or decreases 10 percent for each five-year period in this study.

Figure 13.6 shows the impacts of changes in trip generation rates on the VMT outputs over time. HBWUrban+10% means that all trip generation rates for HBW trips in urban areas have been increased by 10 percent. The pattern or shape is stable across all VMT outputs in this figure. In model year 2000, the largest positive VMT change is 1.16 percent, when the trip generation rates increase 10 percent for HBW trips in rural areas. The smallest negative VMT change is −1.13 percent, when the trip generation rates decrease 10 percent for HBW trips in rural areas.

From model year 2005 to 2030, the largest positive VMT change is 1.12 percent when the trip generation rates increase 10 percent for HBW trips in rural areas at model year 2025; the most negative VMT change is −1.11 percent, when

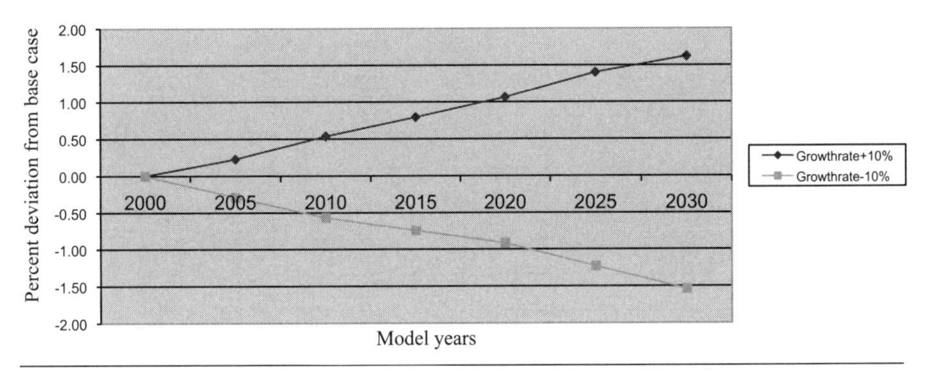

Figure 13.5 Impact of uncertainty in population growth on the VMT outputs

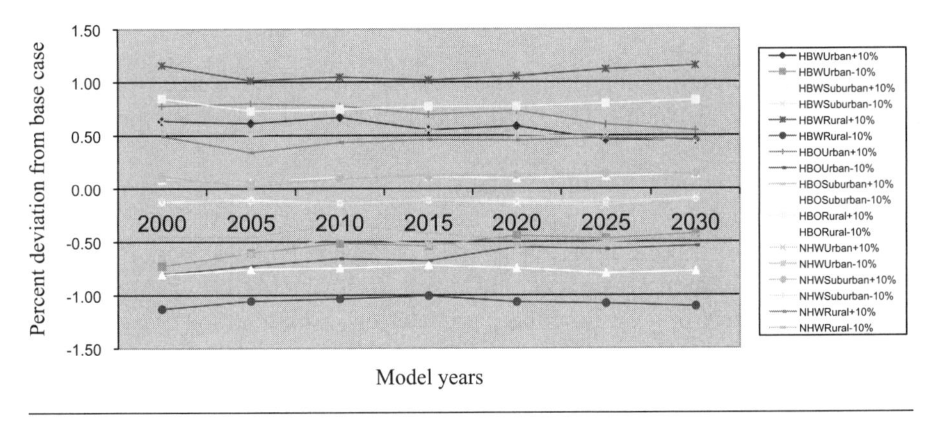

Figure 13.6 Impact of uncertainty in trip generation rates on the VMT outputs

the trip generation rates decrease 10 percent for HBW trips in urban areas at model year 2030. People who live in rural areas generally travel a longer distance between home and work, resulting in the highest sensitivity effects on the VMT outputs.

The impacts of uncertainty in trip attraction rates on the VMT outputs over time are plotted in Figure 13.7. HBW+10% means that all trip attraction rates for HBW trips have been increased by 10 percent. The trip attraction rates have the smallest effect on the VMT outputs in this study. In model year 2000, the largest positive VMT change is 0.0000034 percent, when the trip attraction rates decrease 10 percent for NHW. The most negative VMT change is −0.0000043 percent, when the trip attraction rates decrease 10 percent for HBW. From model year 2005 to 2030, the largest positive VMT change is 0.000025 percent when the trip attraction

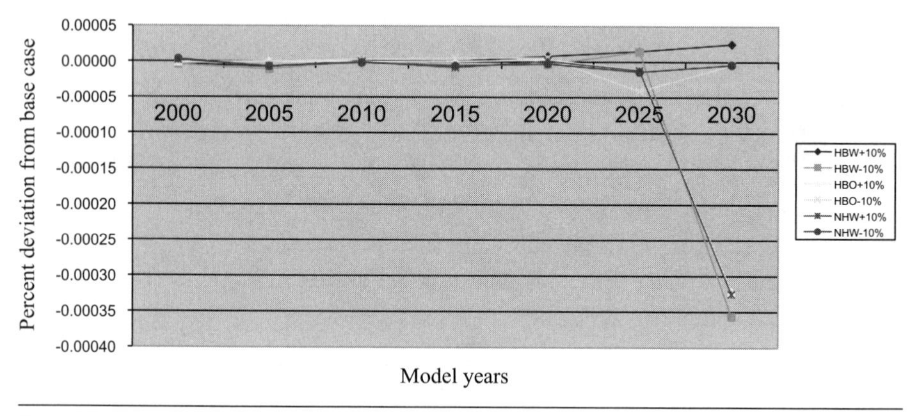

Figure 13.7 Impact of uncertainty in trip attraction rates on the VMT outputs

rates increase 10 percent for HBW. The most negative VMT change is −0.00036 percent at year 2030, when the trip attraction rates decrease 10 percent for HBW at year 2030.

Figure 13.8 presents the impacts of uncertainty in trip distribution parameters on the VMT outputs over time, impacts that are quite constant. HBWGammab+10% means that trip distribution gamma function parameter *b* increases by 10 percent for HBW trips. In model year 2000, the largest positive VMT change is 2.97 percent, when the trip friction function parameter *c* value decreases 10 percent for HBO. The greatest negative VMT change is −2.21 percent, when the trip friction function parameter *c* value increases 10 percent for HBO. From model year 2005 to 2030, the largest positive VMT change is 2.97 percent, when the trip friction function parameter *c* value decreases 10 percent for HBO at year 2010. The most negative VMT change is −2.34 percent, when the trip friction function parameter *c* value increases 10 percent for HBO at year 2005.

Figure 13.9 presents the impacts of uncertainty in trip assignment parameters on the VMT outputs over time. The impacts of uncertainty across model years have high variations over time for trip assignment parameters. Fclass1BPRab+10% means that the values of link performance function parameters *a* and *b* increase by 10 percent for functional class 1. In model year 2000, the largest positive VMT change is 0.12 percent, when the link performance function parameters *a* and *b* decrease by 10 percent for functional class 14. The most negative VMT change is −0.025 percent, when the link performance function parameters *a* and *b* decrease 10 percent for functional class 11. From model year 2005 to 2030, the largest positive VMT change is 0.092 percent, when the link performance function parameters *a* and *b* decrease by 10 percent for functional class 14 at year 2015. The smallest negative VMT change is −0.12 percent, when the link performance

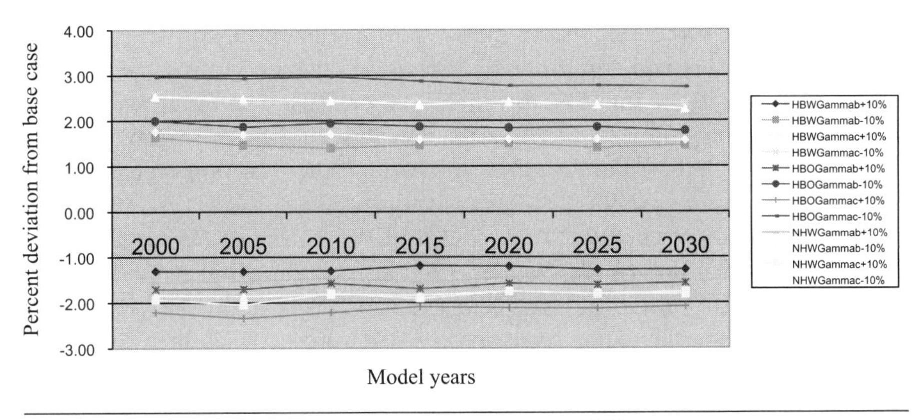

Figure 13.8 Impact of uncertainty in trip distribution on the VMT outputs

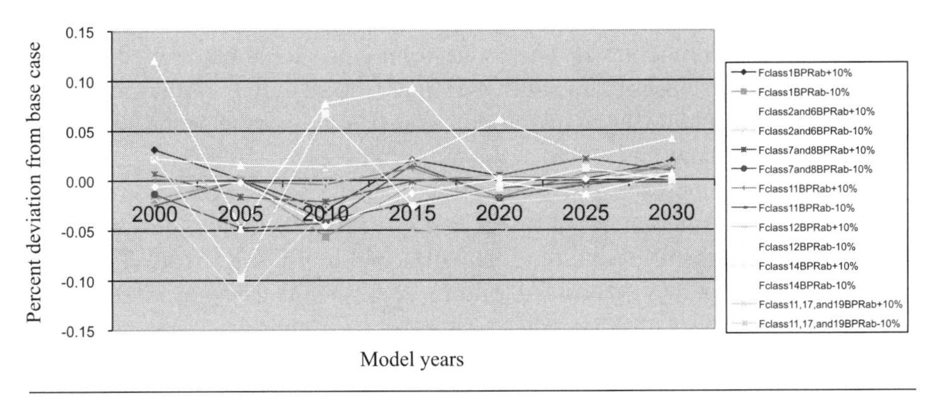

Figure 13.9 Impact of uncertainty in trip assignment on the VMT outputs

function parameters *a* and *b* decrease by 10 percent for functional class 2 and 6 at year 2005.

In all the changes of the VMT outputs, it is found that the most important contributors to uncertainty are trip distribution parameters and trip production rates. Trip attraction rates have the smallest sensitivity influence for the VMT outputs because trip attractions are balanced to trip productions. The VMT outputs are not highly sensitive to changes in population growth rates and trip assignment parameters.

13.5 SUMMARY

This study investigated the sensitivity of the VMT outputs on the model parameters and input data using Indiana's statewide INTRLUDE model system. Results indicate that the VMT outputs are mostly sensitive to trip distribution gamma function parameters and trip production rates. Population growth rates do not have a high sensitivity influence on the VMT outputs. The result agrees with Clay and Johnston (2005). Contrary to the previous study by Krishnamurthy and Kockelman (2003), the exponent of the link performance function is not found to be highly sensitive for the VMT outputs in this study.

The largest change in the VMT outputs in all of the simulation scenarios was 2.97 percent, when the trip distribution friction function parameter *c* decreases 10 percent for HBO at model year 2000. The most negative change in the VMT outputs in all of the simulation scenarios is -2.34 percent, when the trip friction function parameter *c* value increases 10 percent for HBO at model year 2005. All the uncertainty ranges of the VMT outputs are below the ranges of input uncertainty.

This study is valuable for policymakers to understand the impacts of the model parameters and input data on the VMT outputs. Greatest effort and care in assigning values to model parameters and input data should be devoted to variables that cause the most significant change in the model output. Statewide models in other states should be tested on multiple scenarios to investigate the impacts of model parameters and input data on the model outputs.

A univariate approach to the uncertainty analysis was used in this study. The simulations were performed by varying one set of the model parameters or input data by ± 10 percent at a time. The marginal impacts of the VMT outputs were then compared and discussed. Future work should focus on investigating the uncertainty effect of the model parameters and input data simultaneously. Multiple levels of variations should also be considered for future comparisons.

ACKNOWLEDGMENTS

This work was supported by the Joint Transportation Research Program administered by the Indiana Department of Transportation and Purdue University. The contents of this chapter reflect the views of the authors, who are responsible for the facts and the accuracy of the data presented herein and do not necessarily reflect the official views or policies of the Federal Highway Administration and the Indiana Department of Transportation, nor do the contents constitute a standard, specification, or regulation. The authors wish to thank Mr. Steve Smith and Mr. Roy Nunnally of the Indiana Department of Transportation for supporting this study. The efforts of Bernardin, Lochmueller and Associates and Prof. John R. Ottensmann of Indiana University–Purdue University Indianapolis were invaluable in the development of the statewide integrated transportation-land use model system.

REFERENCES

Clay, M. J. and R. A. Johnston. 2005. Univariate uncertainty analysis of an integrated land use and transportation model: MEPLAN. *Transportation Planning and Technology*, 28(3): 149–165.

Indiana Statewide travel demand model upgrade-technical memorandum: Model update and validation. March 2005. Bernardin, Lochmueller, and Associates and Cambridge Systematics.

Krishnamurthy, S. and K. M. Kockelman. 2003. Propagation of uncertainty in transportation land use models: Investigation of DRAM-EMPAL and UTPP predictions in Austin, Texas, Transportation Research Record No. 1831, TRB, National Research Council, Washington DC, 219–229.

Miller, E. J., D. S. Kriger, and J. D. Hunt. 1999. Research and development program for integrated urban models. *Transportation Research Record 1685*, TRB, National Research Council, Washington, DC, 161–170.

Morgan, M. G. and M. Henrion. 1990. *Uncertainty: A Guide to Dealing With Uncertainty in Quantitative Risk and Policy Analysis.* Cambridge, UK: Cambridge University Press.

Ottensmann, J. R. 2003. LUCI land use in central Indiana model and the relationships of public infrastructure to urban development. *Public Works Management & Policy*, 8(1): 62–76.

Ottensmann, J. R. 2007. LUCI2IN Statewide model report on land use classification and model estimation, March 18.

Pradhan, A. and K. M. Kockelman. 2002. Uncertainty propagation in an integrated land use transportation modeling framework: output variation via UrbanSim. *Transportation Research Record 1805*, TRB, National Research Council, Washington, DC, 128–135.

Putman, S. H. 1983. *Integrated Urban Models: Policy Analysis of Transportation and Land Use.* London: Pion.

Waddell, P. 2002. UrbanSim: Modeling urban development for land use, transportation and environmental planning. *Journal of the American Planning Association*, 68(3): 297–314.

Zhao, Y. and K. M. Kockelman. 2002. The propagation of uncertainty through travel demand models: An exploratory analysis. *Annals of Regional Science*, 36: 145–163.

Index

N

P

Q

R

S

T

Z